Thermodynamics of Electrical Processes

MALCOLM McCHESNEY
Senior Lecturer in Thermodynamics
Mechanical Engineering Department
University of Liverpool

WILEY-INTERSCIENCE
a division of John Wiley & Sons Ltd.
London New York Sydney Toronto

Copyright © 1971 John Wiley & Sons Ltd. All Rights Reserved. No part of this publication may be reproduced, stored in a retrieval system, or transmitted, in any form or by any means, electronic, mechanical photocopying, recording or otherwise, without the prior written permission of the Copyright owner.

Library of Congress Catalog Card number 75–166417

ISBN 0 471 58163 1

Made in Great Britain at the Pitman Press, Bath

Preface

I have been teaching engineering thermodynamics to undergraduates for over a decade and in that time student reaction has made it clear to me that the thermodynamic topics traditionally taught to mechanical engineering undergraduates are not necessarily acceptable to students of electrical engineering or physics. All three groups of undergraduates have a similar first year of thermodynamics which usually deals with the basic laws of macroscopic thermodynamics. However, it seems reasonable that the second and any further years of study should apply thermodynamic reasoning to topics of interest to the three different disciplines of mechanical engineering, electrical engineering and physics.

However, an examination of the thermodynamic topics that might be of interest and within the capabilities of electrical engineering and physics students shows that the scope is limited if confined to macroscopic thermodynamics alone. Since it is common practise in North America and Russia and becoming accepted practise in Europe for statistical thermodynamics to find a place in engineering as well as physics undergraduate syllabi, it is sensible to see what simple statistical thermodynamic topics can be dealt with that might interest the student. Accordingly, this book gives a quantitative discussion of the statistical thermodynamic properties of those materials of most interest to the electrical engineer and physicist, mainly metals and semiconductors. In addition the equilibrium approximation is made and the transport properties of these materials are also discussed but only semiquantitatively.

The statistical approach adopted is the simplest, namely the microcanonical ensemble of the thermally perfect gas. Admittedly, for rigour (especially in linking Chapters 1 and 2) the canonical ensemble should have been used; however, I feel that this approach confuses students and therefore I have not used it. My prime objective is to obtain the distribution functions solely in order to develop their applications and I am not concerned in this textbook with formally rigorous developments of the basic postulates.

Although it is becoming the custom to introduce the preliminary statistical concepts by discussing card shuffling and dice throwing, I have avoided this approach since, not being a gambler, I know little of cards or dice. In justification I believe that such introductions are usually either totally ignored or at best only briefly read by students.

I have used the Système International d'Unités (S.I.) units throughout the book because this seems to be the system of units with the greatest chance of widespread adoption, given time. The only concession that I have made to other systems of units is to use the electron volt as an additional unit of energy. I have done this because the joule and its subunits do not seem appropriate for energies measured at the atomic or molecular level. However, in most cases I have quoted energies in both joules (J) and electron volts (eV).

The topics discussed and the formulae obtained are not new—they have all been dealt with before; however, most of these previous discussions have been given by solid-state physicists writing for physics students. I am a professional thermodynamicist and accordingly I have written from this viewpoint since the book is on thermodynamics and not solid-state theory.

It is hoped that some of the contents should appeal to second-year students of electrical engineering and physics who study the atomic theory of matter and yet are obliged to take some thermodynamics as well. In addition, some parts of the book may be of use in final-year undergraduate and introductory post-graduate courses relating to the modern engineering aspects of metals and semiconductors.

It has been said that writing a book is 'transferring bones from one graveyard to another'. Certainly I have read other books in preparing this one, but I would never describe any of these texts as a graveyard. I have been particularly influenced by the excellent texts written by Kittel, Fay, Levy, Lindmayer and Wrigley and Slater. Any errors in this book must, however, remain my sole responsibility.

M. McChesney

Contents

Chapter One: A Brief Review of Macroscopic Engineering Thermodynamics
1.1 Basic definitions of engineering thermodynamics . . . 1
1.2 The properties of a thermodynamic system and its surroundings . 3
1.3 The First Law of Thermodynamics and the property called Energy 5
1.4 The equilibrium of a thermodynamic system 7
1.5 Reversible and irreversible changes of state 9
1.6 The Second Law of Thermodynamics and the property called Entropy 10

Chapter Two: Statistical Thermodynamics of the Thermally Perfect Gas
2.1 The thermally perfect gas 13
2.2 The macrostates of the thermally perfect gas 14
2.3 The quantized translational energy states and levels of the thermally perfect gas; the 'particle in a box' problem . . . 16
2.4 Quantum degeneracy 18
2.5 The microstates of a thermally perfect gas 27
2.6 The Pauli Exclusion Principle 31
2.7 Bose–Einstein statistics 33
2.8 Fermi–Dirac statistics 38
2.9 Maxwell–Boltzmann statistics 41
2.10 The 'preferred' or 'most probable' macrostate 42
2.11 The links between a statistical and a thermodynamic description of a thermally perfect gas 50
2.12 The generalized partition function for Bose–Einstein and Fermi–Dirac statistics 55
2.13 The Maxwell–Boltzmann partition function and the evaluation of β 57
2.14 The macroscopic and microscopic consequences of placing two thermodynamic systems in contact 64

Chapter Three: The Thermally Perfect Maxwell–Boltzmann gas
3.1 The thermodynamic properties of a Maxwell–Boltzmann gas composed of monatomic atoms 69

3.2	The partition function of a particle possessing electronic structure	72
3.3	Thermodynamic properties of an ionized laboratory plasma	74

Chapter Four: The Thermally Perfect Fermi-Dirac Electron Gas in a Metal

4.1	Electron energy bands in a metal (Sommerfeld model)	80
4.2	The thermally perfect degenerate electron gas in a metal	82
4.3	The Fermi level	90
4.4	The actual distribution of electrons in the valence band	92
4.5	Thermodynamic properties of the thermally perfect degenerate electron gas in a metal	96
4.6	Limitations of the thermally perfect degenerate electron gas (Sommerfeld model)	103

Chapter Five: The Thermally Perfect Bose–Einstein Phonon Gas in a Solid

5.1	The thermodynamic properties of a solid	104
5.2	The quantum oscillator	106
5.3	The Debye theory of the lattice	113
5.4	The phonon gas	116
5.5	Asymptotic behaviour of the thermodynamic properties of a phonon gas	123
5.6	The total heat capacity of a solid	130
5.7	The number of phonons in a solid	133

Chapter Six: Transport Properties of Solids

6.1	Transport processes and the equilibrium approximation	137
6.2	Electrical current conduction in metals	139
6.3	The collisional relaxation time and electron-scattering processes	143
6.4	The effect of electron–phonon scattering collisions on the electrical conductivity of the valence-electron gas	151
6.5	A relationship between thermal conductivity and heat capacity	161
6.6	The thermal conductivity of the valence-electron gas	164
6.7	The thermal conductivity of the lattice-phonon gas	170
6.8	The total thermal conductivity of a solid	180
6.9	The Wiedemann–Franz–Lorenz law for electrical conductors	182

Chapter Seven: The Quasifree Thermally Perfect Electron Gas in a Semiconductor

7.1	Electron energy bands in a solid; the valence and conduction bands	190
7.2	The distribution of electron energy states in a band	197

Contents

- 7.3 Carrier concentrations in an intrinsic semiconductor in the classical approximation 200
- 7.4 The Law of Mass Action and the Fermi level in an intrinsic semiconductor 208
- 7.5 Carrier concentrations in an extrinsic semiconductor . . 214
- 7.6 The Fermi level in an extrinsic semiconductor . . . 222
- 7.7 Carrier mobility and carrier scattering processes in a semiconductor 233
- 7.8 The electrical and thermal conductivity of a semiconductor . 241
- 7.9 Diffusion processes in a semiconductor 251

Chapter Eight: Information Theory and Thermodynamics
- 8.1 Information theory and thermodynamics 255
- 8.2 The meaning of Entropy: the Third Law of Thermodynamics . 259
- 8.3 Entropy and Information 262

Appendix 1 Some physical constants in S.I. units . . . 264

Appendix 2 The mole, atomic weight and molar volume . . . 265

Index 269

CHAPTER ONE

A Brief Review of Macroscopic Engineering Thermodynamics

1.1. Basic definitions of engineering thermodynamics

In this chapter we shall give a brief review of elementary engineering thermodynamics, assuming that the reader has already received some teaching in this topic. Thermodynamics is that science dealing with 'the properties of thermodynamic systems and their changes due to heat and/or work and/or mass transfers'.

Let us explain the words used in this definition. A thermodynamic 'system' is some collection of matter occupying some region of space but not necessarily of constant volume. Hence the system is some identifiable collection of matter that we can distinguish from all other matter; it may be a gas, an ionized plasma or more likely, as far as the present book is concerned, a piece of metal or semiconductor. The 'system boundary' is that surface which confines the system, or separates our chosen collection of matter from all other matter. If the system

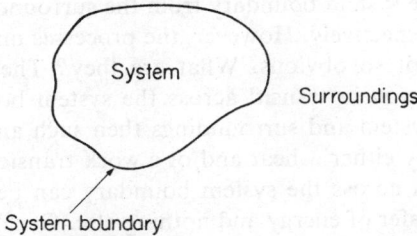

Figure 1.1 The thermodynamic system: the spatial region containing the matter under consideration. The system boundary separates the system from the surroundings which constitute the rest of matter.

is a gas or an ionized plasma, then the system boundary will most likely be some container: if the system is a piece of metal or semiconductor then the

1

system boundaries are the 'edges' of the conductor. To avoid continually using the phrase 'all other matter' we use the alternative 'surroundings' to describe this all other matter. Figure 1.1 shows the three defined quantities.

Let us return to our definition. Thermodynamics is concerned with changes that occur within the system when something happens to change the system. It seems reasonable to assume that, when 'something happens' it likely affects both the system and the surroundings, thus system and surroundings 'interact'. After all, if no interaction occurred then quite possibly both system and surroundings would remain unchanged in space and time and, thermodynamically speaking, this would be a rather dull situation. It would also be rather useless because an engineer always wants to 'use' systems so that they serve him. Systems that remain unchanged and do not interact with their surroundings are called 'isolated systems'.

The idea of an interaction between a system and the surroundings means that the interaction must occur across the system boundary. This raises two questions.

1. What is the nature of the interaction?
2. How do we know if an interaction has occurred? How can we describe the system and its surroundings before and after the interaction in order to find out if any aspect of either the system or the surroundings or both has changed?

Let us answer these questions in turn. The nature of an interaction is such that it represents a transfer process between system and surroundings across the system boundary in either direction. But what is transferred? The answer is that the transfer process can be a heat transfer and/or a work transfer and/or a mass transfer—and nothing else. The idea of a transfer of mass between system and surroundings is clear—at the macroscopic level (i.e. bulk, laboratory level) it means that some collection of matter is either added to or taken from the system by crossing the system boundary from the surroundings which therefore lose or gain matter respectively. However, the processes underlying heat and/or work transfers are not so obvious. What are they? They are no more than manifestations of 'energy in transit' across the system boundary; if energy is exchanged between system and surroundings then such an exchange of energy can only take place by either a heat and/or a work transfer. Therefore, in conclusion an interaction across the system boundary can be either a transfer of of mass and/or a transfer of energy and nothing else. Generally speaking in this book, with one exception, we shall deal with a 'closed' thermodynamic system, that is, one which has a constant mass and therefore allows no mass transfer to or away from it; of course a closed system *will* allow a transfer of heat and/or work.

Now consider the second question raised above. How can we ascertain if anything in the system or surroundings has actually changed if a transfer process has occurred across the system boundary? Clearly we need some variables to describe the system and the surroundings so that before the transfer process occurs we can fully describe the system and the surroundings with these variables. After we suspect a transfer process has occurred, we can re-evaluate these variables and if they have changed then we shall know that a transfer process has actually taken place. But what variables shall we choose? What 'thermodynamic coordinates' shall we use that will adequately describe the system and the surroundings? The appropriate variables are called 'properties' and these are

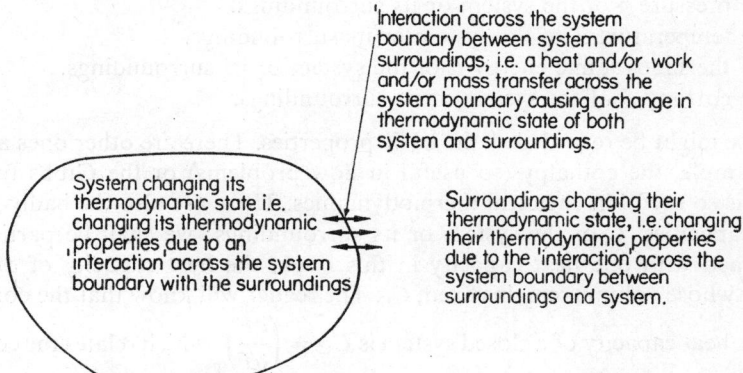

Figure 1.2. A schematic representation of an 'interaction' across the system boundary between a system and its surroundings. The interaction is a transfer process which results in a change of thermodynamic state of both system and surroundings.

such that if we write down the numerical values of an array of properties before a suspected transfer process then we have defined the initial 'thermodynamic state' of the system and the surroundings. Similarly a writing down of the numerical values of the same array of properties after the suspected transfer process defines the final thermodynamic state of the system and surroundings. Figure 1.2 attempts to summarize the idea of the thermodynamic state of a system and its surroundings and the change in the thermodynamic states of system and surroundings due to a transfer process between them.

1.2 The properties of a thermodynamic system and its surroundings

We have now introduced the idea of a thermodynamic property as being a thermodynamic coordinate that describes some aspect of the system or the surroundings. A formal definition of a property is 'any observable macroscopic characteristic of the system or the surroundings which has a single-valued numerical value when the thermodynamic system or surroundings is in an unchanging, or constant, thermodynamic state'. The properties that we shall find useful in this book for describing the system with which we are primarily concerned (usually a piece of metal or a semiconductor) and its surroundings (usually the laboratory in which the piece of metal or semiconductor is situated) are firstly the mass of the system M. Since we shall be largely concerned with closed systems of constant mass, M will be a constant and not a variable. Other properties of the system which may, however, vary are:

1. The volume V of the system or its surroundings.
2. The pressure p of the system or its surroundings.
3. The temperature T of the system or its surroundings.
4. The thermodynamic energy E of the system or its surroundings.
5. The entropy S of the system or its surroundings.

These might be regarded as the basic properties. There are other ones as well; for example, the enthalpy (so useful in flow problems) or the Gibbs function which is so useful in chemical thermodynamics. There are also the badly-named 'heat capacities' C of the system or its surroundings; we shall be particularly concerned with one heat capacity in this book, the heat capacity of a closed system whose volume remains fixed, C_V. The reader will know that the constant-volume heat capacity of a closed system is $C_V = \left(\dfrac{\partial E}{\partial T}\right)_V$ which relates the constant volume heat capacity to the partial differential of the system energy with respect to system temperature.

The concepts of mass, volume, pressure and temperature have long been implanted in our minds from our earliest school days and we shall not formally define them here. Thermodynamics has something specific to say about the definition of temperature which is defined as a corollary of the Zero'th Law of Thermodynamics. What then about the definitions of the other properties, energy E and entropy S of the system or the surroundings? In fact these can be defined in terms of heat and/or work transfers to or from the system, and this is the procedure we shall adopt in this chapter; it is not the only procedure available but is the currently accepted one in engineering thermodynamics.

For a closed system we have the five variables, the properties V, p, T, E, S. Is a heat or work transfer a property? The answer is a definite and unambiguous 'no'! Let us illustrate this point with an example. Consider a piece of extrinsic germanium semiconductor. This is a closed thermodynamic system (of constant mass) since it is an identifiable collection of matter. It has a system boundary along its edges and its surroundings are the laboratory in which it is situated. If there are no interactions across the system boundary, thus no heat and/or work transfers between the semiconductor and the laboratory, then the semiconductor is in a fixed or unchanging thermodynamic state. We can define this thermodynamic state by giving a numerical value to the volume, pressure

A Brief Review of Macroscopic Engineering Thermodynamics

(the concept of a pressure within a solid is discussed in Chapter 4), temperature, energy and entropy of the semiconductor. Does the semiconductor have a numerical value of heat and/or work transfer? The answer is 'no'.

It therefore looks as if the three transfer processes of heat and/or work and/or mass are different from the properties of the system we are primarily considering here. Is there any relationship between them? We might expect there to be a relationship because, after all, as a result of a transfer process, the thermodynamic system and its surroundings change—in particular they change the numerical values of their properties. We might therefore expect *changes* in the thermodynamic properties of a system and its surroundings to be related to the transfer processes. The actual relationships are called the First and Second Laws of Thermodynamics.

1.3 The First Law of Thermodynamics and the property called Energy

Consider a closed thermodynamic system (one of constant mass). The First Law states that

Algebraic sum of the heat and work transfers to and from a system = change of energy of the system

This statement can be regarded as a definition of the change in the property E, the energy of the system. We can frame this statement mathematically using the currently accepted (but rather unfortunate) sign convention for the direction of the transfer processes. Let

$ƌQ$ = positive heat transfer *to* the system from the surroundings
$ƌW$ = positive work transfer *from* the system to the surroundings
dE = infinitesimal increase in the energy of the system due to the heat and work transfers.

Then

$$ƌQ - ƌW = dE \tag{1.1}$$

The reader will note that we have not used the adjective 'infinitesimal' to describe the heat and work transfer processes as we have done for the energy change. This is because heat and work transfers are not variables of the system or its surroundings, thus they are not properties. Consequently they cannot be differentiated as mathematical functions to give infinitesimal increments. We draw a bar through the d symbol, thus ƌ will remind us constantly that heat and work transfers are not properties. However, the five properties V, p, T, E and S are properties, or legitimate variables which are continuous mathematical functions which can be differentiated and give truly infinitesimal increments.

dV, dp, dT, dE and dS, in these properties of the system. This means that we can talk about an infinitesimal change in these properties due to the transfer process but we cannot talk about an infinitesimal change in either heat or work transfer to or from a system. Figure 1.3 attempts to represent equation (1.1) schematically.

Equation (1.1) is known as the differential form of the First Law for a closed system. Let us integrate it from a known initial thermodynamic state of the system before a transfer process occurred to some known final thermodynamic

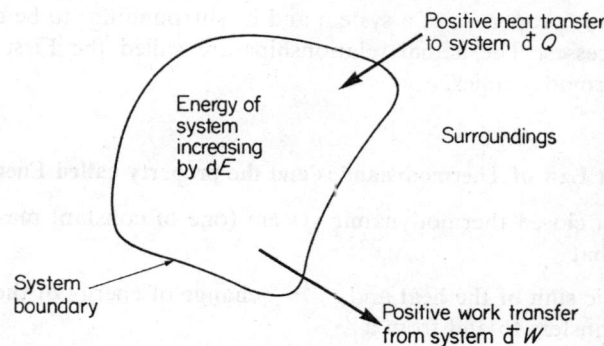

Figure 1.3. Schematic representation of the differential form of the First Law of Thermodynamics for a closed system showing the sign conventions for heat and work transfers which result in an increase of energy of the system.

state of the system after the transfer process has occurred

$$\int_{\text{Initial State}}^{\text{Final State}} \mathrm{d}Q - \int_{\text{Initial State}}^{\text{Final State}} \mathrm{d}W = \int_{\text{Initial State}}^{\text{Final State}} \mathrm{d}E$$

i.e.

$$Q - W = E_{\substack{\text{Final} \\ \text{Thermodynamic} \\ \text{State}}} - E_{\substack{\text{Initial} \\ \text{Thermodynamic} \\ \text{State}}} \qquad (1.2)$$

We note that we *can* write

$$\int_{\text{Initial State}}^{\text{Final State}} \mathrm{d}E = E_{\substack{\text{Final} \\ \text{State}}} - E_{\substack{\text{Initial} \\ \text{State}}} \qquad (1.3)$$

and that we *cannot* write

$$\int_{\text{Initial State}}^{\text{Final State}} \mathrm{d}Q = Q_{\substack{\text{Final} \\ \text{State}}} - Q_{\substack{\text{Initial} \\ \text{State}}} \qquad (1.4)$$

A Brief Review of Macroscopic Engineering Thermodynamics

Equation (1.3) is allowed because E is a property of the system and the system has a unique numerical value of energy E before and after the transfer process. However equation (1.4) is disallowed because if it were true it would imply that Q is likewise a thermodynamic property and that the system has an initial and a final heat transfer—which is meaningless. Likewise an equation similar to equation (1.4) for W is disallowed.

The reader may now be confused as to what distinguishes a property like, for example, E from a non-property such as Q. Let us attempt to remove the confusion by considering the problem of measurement. If we can measure a quantity *within* a thermodynamic system—such as its pressure or its volume or its entropy (never mind how we measure entropy in practise—it can be done indirectly)—then this quantity is a property of the system. If we can only measure a quantity at the *system boundary* (and not *within* the system), then such a quantity is a non-property. Recalling that we have already stated that heat and work transfers are energy in transit we can only identify them at the system boundary where the energy is actually transferred between system and surroundings. Before the interaction occurs, the energy transferred is energy that can be measured either within the system or its surroundings (depending where it is initially located); during the interaction it can only be measured at the system boundary as it traverses between system and surroundings; after the interaction it can be measured once again either within the system or its surroundings (depending on where it has been finally 'deposited').

1.4 The equilibrium of a thermodynamic system

Consider equation (1.1) that describes a closed system. If $đQ = 0$ and $đW = 0$ then $dE = 0$, i.e. from equation (1.2) after integration,

$$E_{\text{Final Thermodynamic State}} = E_{\text{Initial Thermodynamic State}}$$

This is the case where there are no transfer processes occurring between system and surroundings, where the system and surroundings are in *equilibrium* with each other. There are in fact three types of equilibrium between a system and its surroundings.

1. Thermal equilibrium—the temperature is the same throughout the entire volume of the system and equal to the temperature of the surroundings.
2. Mechanical equilibrium—all forces exerted by the system on its surroundings are uniform throughout the entire volume of the system and exactly balanced by the forces exerted by the surroundings on the system.

3. Chemical equilibrium—the chemical composition of the system and its surroundings show no tendency to undergo spontaneous chemical changes and therefore remain unchanged in time.

For an *open* system to be in thermodynamic equilibrium (see below) with its surroundings the above three types of equilibria must all hold with the addition to the criterion of chemical equilibrium that there is no tendency for a spontaneous mass transfer (exchange of particles, molecules or atoms) between system and surroundings.

If all three types of equilibria hold between system and surroundings, then the system is said to be in thermodynamic equilibrium with its surroundings. Such a system, as already remarked, where there are no transfer processes occurring at all and nothing is changing in either system or surroundings, is rather dull and thermodynamically rather useless. For example, our piece of extrinsic germanium sitting on a laboratory table is of not much value to anybody; it has to be used in order to serve us. An engineer always wants to use a thermodynamic system to serve him and the way he does this is to allow an interaction between system and surroundings, thus to allow a transfer process to occur. But such a transfer process must be 'driven'—it will not occur spontaneously. The only way to obtain a 'driving force' that will produce a transfer process is to allow some amount of disequilibrium to occur between system and surroundings. Such a disequilibrium will terminate the state of thermodynamic equilibrium between system and surroundings by causing a transfer process to occur, thereby changing the thermodynamic state of the system and surroundings. Let us give an example. Suppose the system is an ionized laboratory plasma confined in a rigid cylinder fitted with a movable piston. If the plasma pressure exceeds that of the laboratory atmosphere, then the system boundary (the movable piston) will move outwards so as to equalize the pressure between plasma and laboratory atmosphere. This outward movement of the piston means that the surroundings (laboratory atmosphere) will be compressed, as the system does 'displacement work' on the surroundings so that there will be a displacement work transfer of amount W between system and surroundings.

The reader will know that the displacement work done is, in differential form,

$$đW = pdV \qquad (1.5)$$

so that the total displacement work done is

$$W = \int_{\text{Initial Volume of System}}^{\text{Final Volume of System}} pdV \qquad (1.6)$$

Alternatively, if there is a temperature difference between plasma and laboratory atmosphere then there will be a flow of thermal energy (a flow of 'heat'), that is, a heat transfer between plasma and laboratory atmosphere. Similar arguments hold for solids such as our piece of metal or semiconductor. These will expand when heated so that in this case a heat transfer to the solid results in a displacement work transfer by the solid as it pushes back the laboratory atmosphere in order to expand.

1.5 Reversible and irreversible changes of state

Clearly the greater the degree of disequilibrium between system and surroundings, the greater the 'driving force' that will produce a transfer process and the greater the change in the thermodynamic state of the system and its surroundings. In thermodynamics a fundamental distinction arises in discussing the degree of disequilibrium. If the degree of disequilibrium between system and surroundings is no more than infinitesimal, that is if, for example, the difference in pressure or temperature between system and surroundings is no more than truly dp and dT, then as the thermodynamic state of both system and surroundings change due to the work and/or heat-transfer processes, both system and surroundings pass through a sequence of states of thermodynamic equilibrium where every property that describes any state within the sequence is only infinitesimally different from the property describing the immediately adjacent state. This means that the thermodynamic state of both the system and surroundings (the numerical values of the properties describing the system and surroundings) are known *at all times* during the change of state (during the transfer process) as well as the initial and final thermodynamic states before and after the transfer process. Such a change of state proceeding through a sequence of equilibrium states is known as a thermodynamically reversible change of state. If, however, the degree of disequilibrium between system and surroundings is larger than infinitesimal, then once again the thermodynamic state of the system and surroundings will change—but this change will *not* be through a sequence of states of thermodynamic equilibrium; it will be through a sequence of states of non-equilibrium. This means that we will not necessarily be able to write down the numerical values of the properties that describe the system and the surroundings *during* the change of state, that is, during the transfer process. All we can do is to specify the thermodynamic properties of the system and surroundings *before* the transfer process starts (the initial thermodynamic state) and the thermodynamic properties *after* the transfer process has occurred (the final thermodynamic state)—but not in between. Such a change of state proceeding through some sequence of states that are not infinitesimally different from each other, states not of thermodynamic equilibrium but of non-equilibrium between system and surroundings, is known as a thermodynamically irreversible change of thermodynamic state.

Let us give an example of this distinction between reversible and irreversible changes of state. Equation (1.6) states that for a displacement work-transfer process

$$W = \int_{\text{Initial Volume of System}}^{\text{Final Volume of System}} p\,dV$$

Can we perform this integration? The answer viewed purely mathematically will be 'yes' provided we know the p–V relation, the mathematical function $p = p(V)$. From the thermodynamic viewpoint this means that we must know the pressure at every instant during the change of state as the system volume changes. If the change of state is thermodynamically reversible, then we do know $p = p(V)$ at all times. Such a p–V relation is known as 'an equation of state' and we shall be deriving various equations of state in this book. Our ability to specify the equation of state *during* a work-transfer process means that during displacement of the system boundary, it moves back under an infinitesimal force, for instance, infinitesimal pressure difference between system and surroundings and therefore does not accelerate. This is known as a quasi-static change of system volume.

However, if the change of state is irreversible, we cannot perform the integration because we cannot write down the equation of state *during* the change of system volume. Why can we not do so? If, for example, the pressure difference between system and surroundings is more than infinitesimal, we have a finite force pushing back the system boundary. The system boundary therefore accelerates according to Newton's Second Law of Motion. That portion of the system immediately adjacent to the accelerating boundary will try to follow the acceleration, and this could give rise to a distribution of pressures throughout the system—low pressures near the accelerating boundary as the system in that region tries to keep up with the boundary—and higher pressures elsewhere in the system where the effects of the boundary movement have not yet been 'felt'. This means that at any instant of time and therefore at all instants of time during the boundary movement, the system has a determinable volume but does not have a single, fixed pressure because the pressure will vary throughout the volume of system and any measurement that we take of pressure will give a value relative to where the pressure is actually measured within the system. We therefore can neither specify a unique equation of state during the interaction nor integrate equation (1.6). Put succinctly, 'we are always thermodynamically ignorant about the properties of a system during an irreversible change of state'.

1.6 The Second Law of Thermodynamics and the property called Entropy

The Second Law, as stated here, is a statement about a closed system which undergoes a heat transfer, but no work transfer between system and

A Brief Review of Macroscopic Engineering Thermodynamics

surroundings. The heat transfer must be a reversible one which means that there can be no more than an infinitesimal difference in temperature between system and surroundings. The statement is

$$\text{Change (increase) in entropy of a system} = \frac{\text{Reversible heat transfer to the system from the surroundings}}{\text{Temperature of the surroundings}}$$

i.e.

$$dS = \frac{dQ_R}{T} \tag{1.7}$$

where the subscript R mean 'reversible'.

This is shown schematically in Figure 1.4. At first sight this may seem a more restricted type of definition of a property (entropy) than the definition of the other property energy E given by the First Law [equation (1.1)]. However, this is only because equation (1.7) defines entropy in terms of a non-property, by the

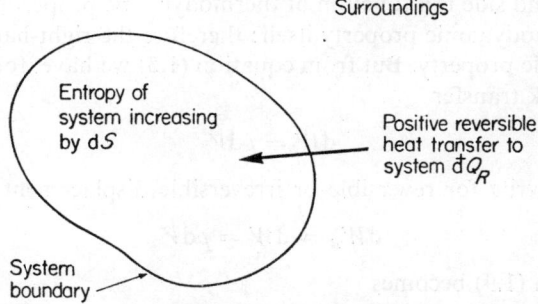

Figure 1.4. Schematic representation of the differential form of the Second Law of Thermodynamics for a closed system showing a reversible heat transfer from the surroundings to the system which results in an increase in entropy of the system.

reversible heat transfer. The reader may ask what happens if the heat transfer is irreversible or if a work transfer occurs as well. The answer to this is to use the above restrictive definition and try to combine it with some more general principle like the First Law and thereby broaden the definition of entropy. It proves useful to define entropy for reversible and irreversible processes in terms of other *properties* rather than non-properties. This can readily be done by combining the First and Second laws thus

$$dS = \frac{dQ_R}{T}$$

$$= \frac{(dW + dE)_R}{T}$$

using equation (1.1). That is

$$TdS = đW_R + dE_R \tag{1.8}$$

However, energy E is a property and the change of energy in any change of state, be it reversible or irreversible, depends upon the initial and final thermodynamic states of the system only [see equation (1.3)] and not upon what happens between these two states. Hence, for all changes of state, reversible or irreversible, between the same initial and final thermodynamic states

$$dE_R \equiv dE$$

so that equation (1.8) becomes

$$TdS = đW_R + dE$$

i.e.

$$TdS - dE = đW_R \tag{1.9}$$

Now the left-hand side is a function of thermodynamic properties and therefore must be a thermodynamic property itself; therefore the right-hand side must be a thermodynamic property. But from equation (1.5) we have, for a reversible or irreversible work transfer

$$đW = pdV$$

Hence we can write for reversible or irreversible displacement work transfers

$$đW_R = đW = pdV$$

so that equation (1.9) becomes

$$TdS = dE + pdV \tag{1.10}$$

or

$$dS = \frac{dE + pdV}{T} \tag{1.11}$$

This gives the infinitesimal change of entropy of a system undergoing either a reversible or irreversible infinitesimal change of state. The integration of equation (1.11) can prove difficult in practise.

This completes our review of the elements of basic engineering thermodynamics, and we shall now turn to examining the thermodynamic system and its surroundings from the microscopic viewpoint in terms of the behaviour of the molecules or atoms comprising both system and surroundings.

CHAPTER TWO

Statistical Thermodynamics of the Thermally Perfect Gas

2.1 The thermally perfect gas

This chapter and, in fact, the rest of the book will be concerned with explaining how the very general thermodynamic concepts briefly reviewed in Chapter 1 can be interpreted and clarified from the molecular or atomic viewpoint. Once we have this 'microscopic' description clear we shall apply it to problems of interest to the electrical engineer, physicist and metallurgist.

At first sight there appears to be very little in common with the molecules of a gas free to move throughout their container and the conduction electrons in a metal or a semiconductor. Surprisingly, as we shall find, we can treat these electrons as an 'electron gas' whose behaviour is similar to the thermally perfect gas much studied by mechanical engineers.

It is convenient and in fact essential at this point to make a clear statement of the type of thermodynamic system we shall be analysing. Any statistical thermodynamic analysis of a system requires us to find the allowed quantized energy states of that system, that is, those states given by proper ('eigen') solutions of the Schrödinger wave equation. In general this is a formidable undertaking, usually involving sophisticated mathematics or extensive numerical calculations. Clearly such complexities are of little concern or interest to the undergraduate. Why these calculations are difficult is that they require consideration of the interactions that occur between the basic microscopic units that make up the system. For example, gas atoms and molecules are known to be surrounded by force fields which cause these particles to attract each other at large separations and repel at close separations; or another example (which we shall discuss later) is a gas of electrons obviously repelling each other electrostatically. Since there are many particles surrounding any particular 'test' particle, each exerting a force on it, we are dealing with what is known as the many-body problem, the mutual interactions between many particles. Such a problem is quite beyond solution in general—even the case of the mutual interaction of only three particles can only be solved under restrictive approximations.

This appears to be a bad start. However, fortunately, there are many thermodynamic systems of great importance to the engineer and physicist where we may either allow for these mutual interactions, in a very simple fashion, or ignore them altogether. Such a simplification allows us to obtain easy and yet accurate theoretical thermodynamic relationships since we can record the total energy of the system as the sum of the energies of the microscopic quantities of which it is composed—whether these be electrons, ions, atoms, molecules, photons, or phonons! This simplification really is a result of assuming that the interaction forces are so small that the average *potential* energy of these forces (such as the spring-like temporary binding between two attracting molecules or the mutual electrostatic repulsive interaction between two electrons) is everywhere and always negligible compared with their average *kinetic* energy.

This type of simplified behaviour was assumed by early workers in kinetic theory over a hundred years ago when they were analysing the perfect or ideal gas so beloved of thermodynamicists. Consequently we shall refer to the approximation of ignoring mutual interactions as the *thermally perfect gas approximation*. *All* of the thermodynamic systems dealt with in this book can be reduced to this thermally perfect gas approximation. Now the reader will know that every gas condenses to a liquid or solidifies at sufficiently low temperatures—and in fact it is just the mutual interaction forces that cause this to happen (the temperature is so low that the potential energy now exceeds the kinetic energy and condensation occurs). Clearly in this case the thermally perfect gas is a hopeless approximation; fortunately such imperfect gas behaviour is of much more interest to the mechanical engineer or physical chemist and we need not consider it in this book. Surprisingly, however, we shall show that we can even analyse some aspects of a solid using the thermally perfect gas approximation; in fact this forms the starting point of our explanation of the electrical and thermal properties of metals and semiconductors.

2.2 The macrostates of the thermally perfect gas

Let us, without further introduction at this point, consider the case of a closed thermodynamic system which is a volume V of a thermally perfect gas at temperature T. From a thermodynamic point of view this system is in a fixed and known thermodynamic state which does not change with time since there are no heat and/or work transfers to or from it and the system is completely described by the thermodynamic properties p, V, T, E, S (although these properties are not necessarily all independent). The gas is composed of N identical or indistinguishable particles, whose positions in space and whose velocities and energies are continually changing. The reader must realise that

Statistical Thermodynamics of the Thermally Perfect Gas 15

even although we are going to assume that these particles do not interact with each other, that is, they are independent particles, they will suffer elastic collisions between themselves and also with the wall of the container; it is just these collisions that are responsible for them gaining or losing velocity and energy where the latter, of course, is conserved. Since one cubic metre of gas at STP contains about 3×10^{25} particles we cannot and need not keep track of them individually (see Appendix 1 for numerical constants). What we shall try to do, however, is to consider groups of particles where each group will contain a large number of particles such that every particle in any group is characterized by certain common features.

Clearly two features common to all of the particles in all of their groups is their mass and their indistinguishability from each other. In addition, quantum theory states that at any one instant of time and at all subsequent instants of time each and every particle can only possess a range of discrete energies ε_0, ε_1, ε_2 $\varepsilon_3 \ldots \varepsilon_n \ldots$, and no energy in between. We should note that these allowed energies of any particle contain no contribution from interactions with any other particle—we have already stated that such *external* potential energies are negligible in our thermally perfect gas approximation. Hence, all of the energies ε are either purely translational, that is kinetic if each particle is structureless (we shall examine the meaning of this in Chapter 3) or the energies are both translational, that is kinetic and *internal* potential energies if the particles have structure. Hence, at *any* instant of time,

$$\left.\begin{array}{l} N_0 \text{ particles will possess energy } \varepsilon_0 \\ N_1 \text{ particles will possess energy } \varepsilon_1 \\ N_2 \text{ particles will possess energy } \varepsilon_2 \\ \quad \text{etc.} \\ N_n \text{ particles will possess energy } \varepsilon_n \\ \quad \text{etc.} \end{array}\right\} \quad (2.1)$$

This is known as a *macrostate* of the gas. Although we have disallowed particle energies lying between ε_0, ε_1, ε_2, ε_3 ... we have not imposed any limit on the size of the energy gap between these neighbouring quantum energy levels.

Due to particle random motion and elastic collisions each and every particle will be constantly changing its energy and the N_n may be changing so that the gas passes through many (about 10^{29}) macrostates per second.

Our problem is to determine the nature of these macrostates; in particular we want to find formulae for the N_n and ε_n and also ascertain if any one (or any set) of macrostates appears more often than any other. The reason is that if, as the particles randomly change their macrostate description, one macrostate occurs much more often than any other, then this preferred macrostate might be linked with the thermodynamic state which, as we have seen, is not changing with time.

It should be noted that each and every macrostate is subject to two restrictions

$$\sum_n N_n = N \qquad (2.2)$$

This merely states that the number of particles in each and every quantum energy level adds up to the total number of particles in the container. In addition, we have the thermally perfect gas approximation

$$\sum_n N_n \varepsilon_n = E \qquad (2.3)$$

i.e. the sum of the individual allowed energies equals the (thermodynamic) energy of the system. This latter equation is a very important one because it forms our first link between the microscopic description of the system (the left-hand side containing the N_n and the ε_n) and the macroscopic or bulk description of the system (the right-hand side containing the thermodynamic property, energy E).

2.3 The quantized translational energy states and levels of the thermally perfect gas; the 'particle in a box' problem

Ignoring for the time being, the internal potential energies of the particles, let us now determine the ε_n, the allowed quantized translational energies of the individual particles as they move in their container. These are determined from the Schrödinger wave equation which for any one particle in the container can be written

$$\nabla^2 \psi + \frac{2m}{\hbar^2}(\varepsilon - \bar{V})\psi = 0$$

his is the time independent form where

ψ = matter wave-amplitude function
$\bar{V}$ = potential energy of the particle
  = 0 for the thermally perfect gas approximation
m = mass of the particle
$\hbar$ = Planck's constant h divided by 2π

In general this wave equation has an infinite number of solutions. However, the proper solutions (those allowed) are those that satisfy the boundary conditions which are that ψ is single-valued, finite, continuous and vanishing at the

Statistical Thermodynamics of the Thermally Perfect Gas

walls of the container. This latter condition follows since if the particle is to remain in the container ψ must be zero everywhere outside the container—and since ψ is continuous it must be zero at the walls of the container.

Suppose our container is situated with one corner at the origin of a three-dimensional cartesian coordinate system and such that the dimensions of the container are given by

$$x = 0 \text{ to } x = a; y = 0 \text{ to } y = b, z = 0 \text{ to } z = c$$

The wave equation is for this 'particle in a box'

$$\frac{\partial^2 \psi}{\partial x^2} + \frac{\partial^2 \psi}{\partial y^2} + \frac{\partial^2 \psi}{\partial z^2} + \frac{2m}{\hbar^2} \varepsilon \psi = 0 \quad (2.4)$$

Since the motion of a particle in any one cartesian direction is independent of its motion in the other two directions, let us look for solutions which are separable in x, y, z. i.e.

$$\varepsilon = \varepsilon_x + \varepsilon_y + \varepsilon_z \quad (2.5)$$

Hence, for example, we will have the equation,

$$\frac{\partial^2 \psi_x}{\partial x^2} + \frac{2m}{\hbar^2} \varepsilon_x \psi_x = 0$$

with equivalent equations for the y and z directions.

The solution for the x-direction is,

$$\psi_x = A \sin \left(\frac{2m\varepsilon_x}{\hbar^2}\right)^{\frac{1}{2}} x + B \cos \left(\frac{2m\varepsilon_x}{\hbar^2}\right)^{\frac{1}{2}} x$$

Since $\psi_x = 0$ at $x = 0$ and $x = a$ then $B = 0$. However, $\psi_x = 0$ at $x = a$ cannot be satisfied by choosing some value for A; it can only be satisfied by assuming

$$\left(\frac{2m\varepsilon_x}{\hbar^2}\right)^{\frac{1}{2}} a = \text{multiple of } \pi$$

$$= n_x \pi \quad (2.6)$$

where n_x is a positive integer (which must be non-zero since if $n_x = 0$ then $\psi_x = 0$ and there would be no particle anywhere!). n_x is called a translational (energy) quantum number. Hence

$$\varepsilon_{n_x} = \frac{\hbar^2 \pi^2}{2ma^2} \cdot n_x^2 = \frac{h^2}{8m} \cdot \frac{n_x^2}{a^2} \quad (2.7)$$

These are the eigenvalues, the allowed quantized translational energies of the particle. No other particle translational energies are allowed; this quantization of the translational energy arises solely from the boundary conditions, due to the fact that the particle *is* in a container. If the particle had been in 'infinite' space it would not have had a quantized translational energy. Because of equation (2.5) we can write for the full three-dimensional solution

$$\varepsilon_{n_x,n_y,n_z} = \frac{h^2}{8m}\left(\frac{n_x^2}{a^2} + \frac{n_y^2}{b^2} + \frac{n_z^2}{c^2}\right) \tag{2.8}$$

We have thus solved one part of the problem of describing a macrostate of the gas as we have found a formula for the ε_n of equation (2.1). For simplicity, and without any loss of generality, let us assume that the container is a cube of side L, then the equation above becomes

$$\varepsilon_{n_x,n_y,n_z} = \frac{h^2}{8mL^2}(n_x^2 + n_y^2 + n_z^2)$$

i.e.

$$\varepsilon_n = \frac{h^2}{8mL^2}(n_x^2 + n_y^2 + n_z^2) \tag{2.9}$$

i.e.

$$\varepsilon_n = \frac{h^2}{8mV^{\frac{2}{3}}}(n_x^2 + n_y^2 + n_z^2)$$

where $n^2 = n_x^2 + n_y^2 + n_z^2$ and V is the volume of the container.

2.4 Quantum degeneracy

Although each choice of a set of quantum numbers n_x, n_y, n_z corresponds to a different ψ since the solution of equation (2.4) is

$$\psi_{n_x,n_y,n_z} = \psi_n = A \sin\left(\frac{\pi n_x x}{a}\right) \sin\left(\frac{\pi n_y y}{b}\right) \sin\left(\frac{\pi n_z z}{c}\right) \tag{2.10}$$

we see that it does not necessarily give a different translational energy ε_n. This is a vitally important point, whilst each set of translational quantum numbers specifies a unique wave function and a unique *translational energy state*, we see from equation (2.10), it does not necessarily specify a unique *translational energy* as given by equation (2.9). For example, consider Table 2.1 below.

We see from Table 2.1 and equations (2.9) and (2.10) that *different* sets of translational quantum numbers can give the *same* value of the translational

Statistical Thermodynamics of the Thermally Perfect Gas

Table 2.1 Energy-state degeneracy of the translational energy states of a particle moving randomly in a cubical container. The third column gives the statistical weight or degeneracy of the energy level, whose energy is given by the second column.

Translational energy-state description; value of the translational quantum numbers			Value of $n^2 = n_x^2 + n_y^2 + n_z^2$. Also, value of the translational energy of the state in units of $h^2(8\,mV^{2/3})^{-1}$	Number of degenerate translational energy states; number of coincident translational energy states, all of the same translational energy but different wave functions. This is the degeneracy g_n of the translational energy level $\varepsilon_{n_x, n_y, n_z}$
n_x	n_y	n_z		
1	1	1	3	1
2	1	1		
1	2	1	6	3
1	1	2		
2	2	1		
2	1	2	9	3
1	2	2		
2	2	2	12	1
3	1	1		
1	3	1	11	3
1	1	3		
3	2	1		
3	1	2		
2	3	1	14	6
2	1	3		
1	3	2		
1	2	3		
3	2	2		
2	3	2	17	3
2	2	3		
4	1	1		
1	4	1	18	3
1	1	4		
etc.			etc.	etc.

energy. For example, consider the three translational energy states characterized by

$$n_x = 2 \quad n_y = 1 \quad n_z = 1 \quad \text{State 1}$$
$$n_x = 1 \quad n_y = 2 \quad n_z = 1 \quad \text{State 2}$$
$$n_x = 1 \quad n_y = 1 \quad n_z = 2 \quad \text{State 3}$$

This set of translational quantum numbers gives three wave functions ψ_n

$$\psi_{2,1,1}, \psi_{1,2,1} \text{ and } \psi_{1,1,2}$$

and we note by substituting the n_x, n_y, n_z values into equation (2.10) that these three wave functions are not equal, i.e.

$$\psi_{n=\sqrt{6}} = \psi_{2,1,1} \neq \psi_{1,2,1} \neq \psi_{1,1,2}$$

However, if we substitute the n_x, n_y, n_z values into equation (2.9) then we get the *same* energy, i.e.

$$\varepsilon_{n=\sqrt{6}} = \varepsilon_{2,1,1} = \varepsilon_{1,2,1} = \varepsilon_{1,1,2} = \frac{h^2}{8mL^2} \cdot (6)$$

This means the translational energy states are not distinguishable from each other by the value of their energy alone; this is known as translational energy-state degeneracy. How then do we distinguish translational energy states? The answer is that it is not sufficient to merely quote their energy ε_n; what we must quote is their complete set of translational quantum numbers n_x, n_y, n_z which give unique wave functions ψ_n. In general we say that any translational energy *level* of energy ε_n has a degeneracy or statistical weight g_n and we take this to mean that that translational energy level is made up of g_n translational energy *states*. All of these translational energy states have different wave functions but the same translational energy.

In our analysis we shall refer to both translational energy states and translational energy levels, depending on the problem under discussion. The reader should remember that, by definition:

a. A translational energy *state* of energy ε_n has a degeneracy or statistical weight of 1 and is uniquely fixed once n_x, n_y, n_z are given.
b. A translational energy *level* of energy ε_n may have a degeneracy or statistical weight of g_n and is not uniquely fixed by a given set of n_x, n_y, n_z. In the above numerical problem we say that the energy states associated with $\psi_{2,1,1}$, $\psi_{1,2,1}, \psi_{1,1,2}$ all have unit degeneracy but that the energy level $\varepsilon_{n=\sqrt{6}}$ has a degeneracy of 3.

It is just translational energy-level degeneracy that is responsible for the enormous difference in behaviour between a thermally perfect gas of

Statistical Thermodynamics of the Thermally Perfect Gas

molecules in a container and a thermally perfect gas of free electrons in a metal. The reason is that in our description of a macrostate we allocated N_n particles to translational energy level ε_n. However, we must now allow for the fact that this energy level may be g_n-fold degenerate comprised of g_n different translational energy states. Although these translational energy states have the same energy they are different states (because they have different wave functions), and therefore moving particles from one state to another (both of the same energy) constitutes a new allocation. Before discussing the effect of translational energy degeneracy on our description of the macrostate, let us further consider the calculation of translational energy-level degeneracy as given in Table 2.1.

Since energy-level degeneracy is so important and, as we shall see in our later analysis, we shall be dealing with many millions of translational energy levels, it would be quite impossible to use a trial-and-error method such as in Table 2.1 to determine the degeneracy of translational energy levels. In fact we can do this quite readily since we can derive a formula that gives the number of translational energy states for a particle of known energy ε. Consider equation (2.9) which we may rewrite as

$$n_x^2 + n_y^2 + n_z^2 = \frac{8mV^{\frac{2}{3}}\varepsilon}{h^2}$$

This is the equation of a sphere in a coordinate space $0n_x$, $0n_y$, $0n_z$ and of radius equal to the square root of the right-hand side, i.e. radius $\left(\dfrac{8mV^{\frac{2}{3}}\varepsilon}{h^2}\right)^{\frac{1}{2}}$.

Since the translational quantum numbers n must be positive integers, we need only consider the positive octant of this sphere which we can draw as in Figure 2.1.

The allowed translational energy states can be represented by points in this 'three-dimensional quantum number space'. If we imagine dividing the volume into small cubes of unit volume then there will be one point (one allowed translational energy state) in the volume of each small cube.

Provided our spherical octant is 'not too small' then we can say that for our particle

Number of translational energy states between energy 0 and energy ε

$$= \Gamma(\varepsilon)$$
$$= \text{volume of the octant}$$
$$= \tfrac{1}{8}(\tfrac{4}{3}\pi)\,(\text{Radius})^3$$
$$= \frac{1}{8}\frac{4\pi}{3}[(8mV^{\frac{2}{3}}\varepsilon h^{-2})^{\frac{1}{2}}]^3$$

i.e.

$$\Gamma(\varepsilon) = \frac{4\pi}{3}(2m\varepsilon)^{\frac{3}{2}}Vh^{-3} \qquad (2.11)$$

To give the reader some idea of the numerical size of this quantity and of the value of the translational quantum numbers that describe translational energy states, let us consider a helium atom of mass $6\cdot 7 \times 10^{-27}$ kg in a container of one litre such that the atom is moving with a velocity of 10^3 m/s, this being a

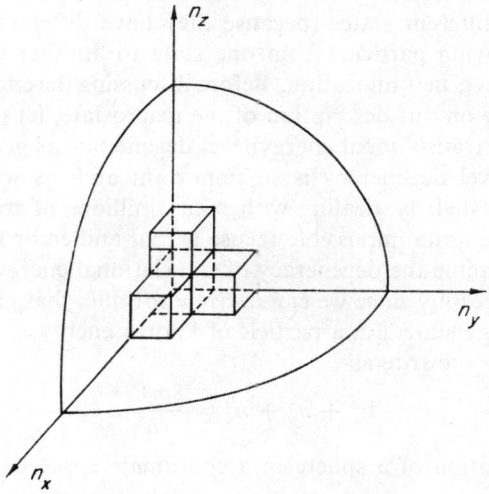

Figure 2.1. The allowed translational energy states of a single particle moving at random in a cubical container. Each allowed state is represented by the corner 'points' of a simple cubical framework in positive quantum number space whose axes are $0n_x$, $0n_y$, $0n_z$.

reasonable value for a helium atom if it were in a container full of helium gas at STP.

Translational energy of the helium atom

$$= \tfrac{1}{2}mv^2$$
$$= \tfrac{1}{2}(6\cdot 7 \times 10^{-27}) \times (10^3)^2$$
$$= 3\cdot 35 \times 10^{-21} \text{ J}$$

Now,

$$n^2 = n_x^2 + n_y^2 + n_z^2$$
$$= 8mV^{\tfrac{2}{3}}\varepsilon h^{-2}$$
$$= \frac{8 \times 6\cdot 7 \times 10^{-27} \times (0\cdot 1)^{\tfrac{2}{3}} \times 3\cdot 4 \times 10^{-21}}{(6\cdot 6 \times 10^{-34})^2}$$

whence

$$n \sim 10^9 \qquad\qquad (2.12)$$

Statistical Thermodynamics of the Thermally Perfect Gas

We see that our helium atom is characterized by a very high translational quantum number. Using equation (2.11), we can further deduce that

$$\Gamma(\varepsilon) \sim 10^{28} \tag{2.13}$$

i.e. there is an enormous number of translational energy states lying between energies of (essentially) zero and $3\cdot 35 \times 10^{-21}$ J.

Equation (2.11) can be used to tell us something about translational degeneracy because we can write

Number of translational energy states between energy ε and energy $\varepsilon + \mathrm{d}\varepsilon$

$$= g(\varepsilon)\mathrm{d}\varepsilon$$

$$= \frac{\mathrm{d}}{\mathrm{d}\varepsilon}\Gamma(\varepsilon)$$

i.e.

$$g(\varepsilon)\mathrm{d}\varepsilon = 2\pi(2m)^{\frac{3}{2}}Vh^{-3}\varepsilon^{\frac{1}{2}}\mathrm{d}\varepsilon \tag{2.14}$$

where $g(\varepsilon)$ is known as the 'density of states' functions.

It gives us the number of translational energy states within a given energy range $\mathrm{d}\varepsilon$ and if $\mathrm{d}\varepsilon$ is taken small enough we may say that all the states in the group have the same energy. We see that this number increases parabolically with the square root of the energy, thus the degeneracy of translational quantum levels depends parabolically on their energy. We shall require these formulae when we come to discuss Fermi–Dirac statistics. In addition, and more relevantly for our immediate purpose, we can write

Average energy separation between successive translational energy levels

$$= \frac{1}{g(\varepsilon)}$$

$$= \frac{h^3}{2\pi(2m)^{\frac{3}{2}}V\varepsilon^{\frac{1}{2}}} \tag{2.15}$$

Consider the helium atom discussed above. Equation (2.15) shows that the average energy separation between translational levels is about 10^{-48} J which is very, very small compared with the translational energy of the atom itself which is $3\cdot 4 \times 10^{-21}$ J. The fact that this spacing between successive translational energy levels is very, very small might have been expected. The reader will likely be aware of the formula for the Maxwellian distribution of velocities (or translational energies) of the particles of a gas and the fact that this formula does not contain Planck's constant. The reason is that in the derivation of this formula and also in the derivation of equations (2.14) and (2.15) we have *assumed* that the translational energy ε is a *continuous* mathematical function because, after all, we have differentiated it. We can see the justification for this because equation (2.15) shows us that the 'scale' of translational energy quantization is so fine that it can be ignored for many purposes, so we can say that the particles have an essentially continuous distribution of translational energies. Despite this conclusion it is still sensible to continue considering the effects of quantization in discussing our macrostate. The reason is that although we have shown that separations between translational energy levels are exceedingly narrow, so far we have not commented on whether there is a restriction on the number of particles allowed to occupy any particular translational energy state or level. Since there is in fact such a restriction in some cases (see Section 2.6 below) we shall not abandon quantum detail at this stage. Furthermore energy quantization is required for particles possessing internal structure as in Chapter 3.

We can now draw a picture of the translational energy *levels* of a single particle in a container; they form an array whose translational energies are such that

$$\varepsilon_0 < \varepsilon_1 < \varepsilon_2 < \varepsilon_3 \ldots$$

Each of these translational energy levels is likely degenerate and composed of many translational energy states. We can picture this array of allowed translational energy levels as a vertical 'ladder' with each 'rung' representing one of the allowed translational energies ε; Figure 2.2a illustrates this ladder. It must be stressed that Figure 2.2a is *not* a graph of translational energy versus some other quantity; it is simply a vertical axis showing the numerical values of the allowed translational energy levels. Each horizontal 'rung' of this energy 'ladder' represents an allowed translational energy. Since each allowed translational energy is described by the quantum number integers n_x, n_y, n_z and since there is no limit to the numerical values that these may take, there will be an infinite number of 'rungs' on the 'ladder', i.e. an infinity of allowed translational energy levels. Of course we have *not* said, so far, if any particular level is actually occupied by one or more particles; all we have done is to discuss the array of allowed translational energy levels regardless of whether they are occupied. Each 'rung' of the 'ladder', each energy level of the array, is such that the energy of any particular level is greater than the energy of the level immediately below it. Furthermore, equation (2.15) shows that the energy separation between these adjacent, successive levels, is proportional to $\varepsilon^{-\frac{1}{2}}$ so that as the numerical value of the energy of the levels increases, their energy separation decreases. In other words the 'rungs' of the translational 'ladder' are not equidistant but get closer and closer together the higher the energy of the level.

Since the N particles of our thermally perfect gas in a container are independent, i.e. do not interact (other than at the instant of collision), the allowed translational energy levels of each of the N particles considered both *separately*

Statistical Thermodynamics of the Thermally Perfect Gas

and *together* are as shown in Figure 2.2(a). We can therefore use the array of translational energy levels shown in Figure 2.2(a) to give a picture of a macrostate at any instant of time. Figure 2.2(b) shows such a macrostate; each dot represents a particle and the number of dots on a particular 'rung' represents the 'occupancy' of that level. Of course only simple numbers of particles are

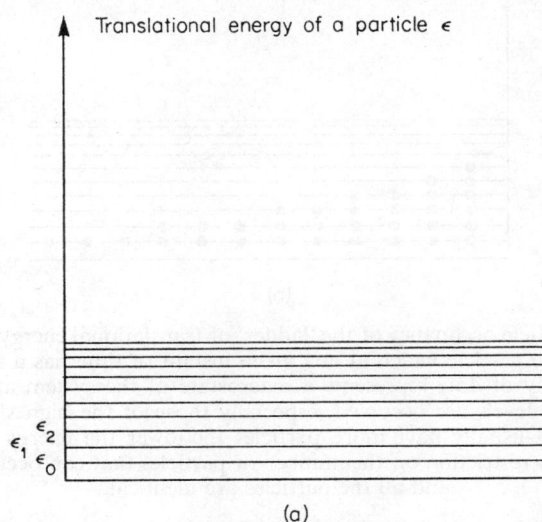

(a)

Figure 2.2(a). Schematic representation of the 'ladder' of allowed translational energy levels of a single particle moving randomly in a container; the energy spacing between adjacent levels decreases as the energy of the level increases. There is no limit to the number of allowed levels which therefore extend indefinitely up the energy scale getting closer and closer together.

shown in Figure 2.2(b) occupying the various energy levels; in practice these numbers can be exceedingly large.

We must now extend our definition of the macrostate to allow for translational energy-level degeneracy. Whereas before we said that, for example, N_n particles were contained in the energy level ε_n we must now recognize that this level may be composed of many coincident translational energy states. We can represent the degeneracy of the allowed energy levels (where the levels are shown in Figure 2.2(a)) as shown in Figure 2.3. Here in the mind's eye we subdivide each 'rung' of the 'ladder' of energy levels into 'part-rungs' where each 'part-rung' can be thought of as a distinct translational energy *state*—whilst the entire 'rung' is the translational energy *level*. In Figure 2.3 we have taken any one of the infinite array of translational energy levels of Figure 2.2, say energy level ε_n and subdivided it into energy states. Figure 2.3 shows eight particles all of the same translational energy ε_n, all occupying the same energy

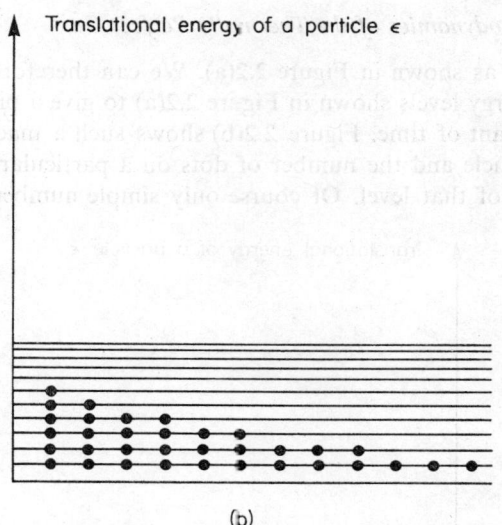

Figure 2.2(b). Particle occupancy of the 'ladder' of translational energy levels. Each dot represents a single particle which at any given instant of time has a specified energy; therefore the array of dots represents a macrostate of the system at that instant of time. Not all the levels are occupied, especially those of the highest energies; those that are occupied usually have more particles the lower the energy of the level. As drawn, there is no restriction on the number of particles that can occupy a given level and all the particles are identical.

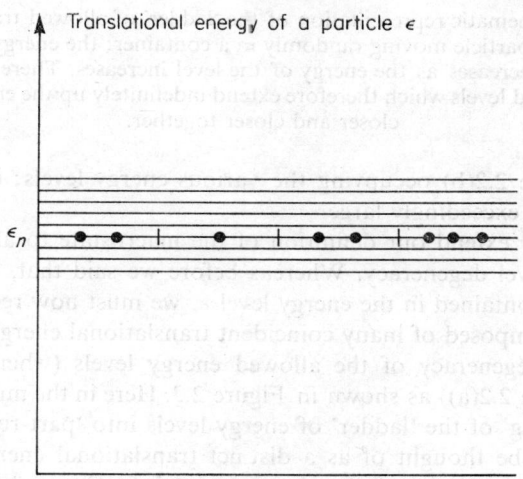

Figure 2.3. The effect of translational energy level degeneracy on the allowed translational energy level ε_n. The level is shown subdivided into four separate energy states all having the same energy ε_n. Each of these energy states contains 2, 1, 2, 3, identical particles respectively.

Statistical Thermodynamics of the Thermally Perfect Gas

level ε_n and distributed amongst the four energy states (all of energy ε_n) that comprise the energy level ε_n. As already noted, we give the alternative names 'statistical weight' or 'degeneracy' to the number of 'part-rungs' into which any 'rung' is subdivided that is, the number of coincident energy states in any energy level. The symbol that we have introduced for the statistical weight of a level is g so that we can say, for Figure 2.3, that we have $g_n = 4$ for the translational energy level ε_n whilst $g_n = 1$ for each of the four translational energy states of that level.

2.5 The microstates of the thermally perfect gas

To see the effect of translational energy-level degeneracy on macrostates let us consider a very simple (but rather physically unlikely) analysis which illustrates the essential features. Consider a thermodynamic system which is a container with only three identical independent particles in it. Let the allowed translational energies of these three particles be

$$\varepsilon_0 = 0; \; \varepsilon_1 = \varepsilon; \; \varepsilon_2 = 2\varepsilon; \; \varepsilon_3 = 3\varepsilon \ldots \text{etc.}$$

as determined by the Schrödinger wave equation and its solution given in equations (2.4), (2.9) and (2.10). Let us suppose that each of these translational energy levels is two-fold degenerate and that the total translational energy of the system is $E = 3\varepsilon$. Let us enumerate the macrostates and microstates of such a system.

Consider macrostate 1 characterized by $N_0 = 1$; $N_1 = 1$; $N_2 = 1$; $N_3 = 0$; we see that because of translational energy level degeneracy and the fact that we can arrange the particles in either of the two 'part-rungs' in each energy-level 'rung', there are no less than eight microstates appropriate to macrostate 1 where a microstate is the name given to a particular arrangement of particles among the degenerate translational energy levels—see Figure 2.4. So that the reader is clear on the distinction between macrostate and microstate we can say that a macrostate is an allocation of particles amongst the various energy *levels* without considering the 'fine' detail of their degeneracy, whilst a microstate is an allocation amongst the various energy *states* (where of course energy levels are made up of degenerate energy states). Clearly each microstate must obey the overall requirements of the macrostate to which it belongs and as our example (Figure 2.4) shows there may be many microstates producing the same macrostate. To summarize we note that

a. Every microstate obeys the macrostate requirement.

$$N_0 = 1 \quad N_1 = 1 \quad N_2 = 1 \quad N_3 = N_4 = \ldots = 0$$

b. Every microstate obeys equation (2.2)

$$\sum_n N_n = N$$

i.e.

$$N_0 + N_1 + N_2 = 3$$

c. Every microstate obeys equation (2.3)

$$\sum_n \varepsilon_n N_n = E$$

i.e.

$$N_0\varepsilon_0 + N_1\varepsilon_1 + N_2\varepsilon_2 = N_0 \cdot 0 + N_1 \cdot \varepsilon + N_2 \cdot 2\varepsilon$$
$$= 1 \cdot 0 + 1 \cdot \varepsilon + 1 \cdot 2\varepsilon$$
$$= 3\varepsilon = E$$

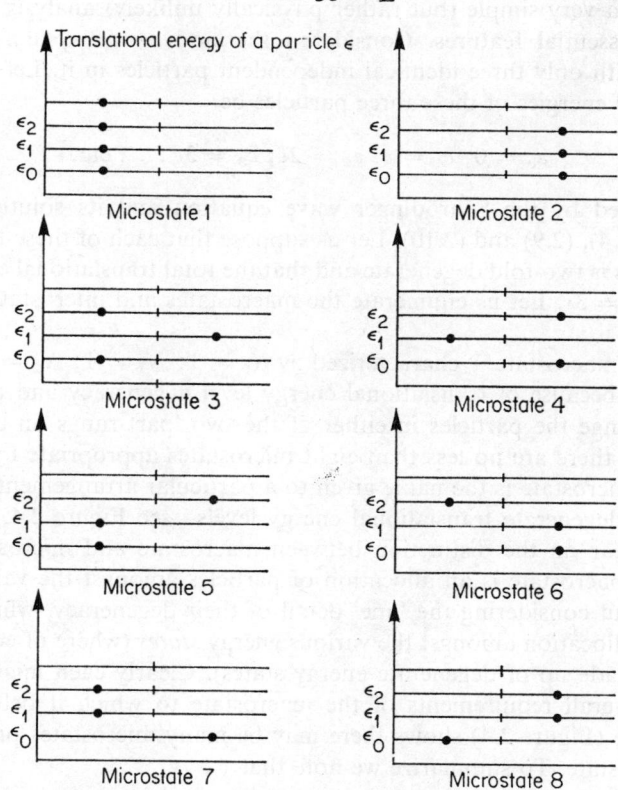

Figure 2.4. The eight microstates of macrostate 1, which arise as a consequence of the effect of translational energy level degeneracy of the system comprising three identical particles to be allocated to translational energy levels of two-fold degeneracy such that the total energy of the system is 3ε.

Statistical Thermodynamics of the Thermally Perfect Gas

However, this is not the whole story; we can still satisfy equations (2.2) and (2.3) by considering another macrostate,

Macrostate 2 $N_0 = 2; N_1 = 0; N_2 = 0; N_3 = 1; N_4 = N_5 = \ldots = 0.$

Instead of drawing the various microstates for this macrostate, we shall merely list them in Table 2.2.

Table 2.2 The six microstates of macrostate 2.

	Energy level ε_0	Energy level ε_1	Energy level ε_2	Energy level ε_3	Energy level ε_4
Microstate 1	2, 0	0, 0	0, 0	1, 0	0, 0
Microstate 2	2, 0	0, 0	0, 0	0, 1	0, 0
Microstate 3	0, 2	0, 0	0, 0	1, 0	0, 0
Microstate 4	0, 2	0, 0	0, 0	0, 1	0, 0
Microstate 5	1, 1	0, 0	0, 0	1, 0	0, 0
Microstate 6	1, 1	0, 0	0, 0	0, 1	0, 0

where, for example, 2, 0 means that there are two particles in the first 'part-rung' of the 'rung' which is energy level ε and none in the second 'part-rung' of the same energy level. We see that there are six microstates compatible with the requirements of macrostate 2. This, however, is not all. There is yet another macrostate with its associated microstates that also satisfies equations (2.2) and (2.3). This is

Macrostate 3 $N_0 = 0; N_1 = 3; N_2 = 0$ etc.

By reasoning similar to that above we can readily show that there are four microstates for this macrostate, and these are given in Table 2.3.

Table 2.3 The four microstates of macrostate 3.

	Energy level ε_0	Energy level ε_1	Energy level ε_2
Microstate 1	0, 0	3, 0	0, 0
Microstate 2	0, 0	2, 1	0, 0
Microstate 3	0, 0	1, 2	0, 0
Microstate 4	0, 0	0, 3	0, 0

Let ω_i = number of microstates associated with any particular macrostate i. Then we can summarize our problem and its answer as

Translational energy levels of system = $n\varepsilon$, where $n = 0, 1, 2, 3, \ldots$
Degeneracy of each level $n = g_n = 2$ for all values of n
Total number of particles in system = $\sum\limits_{n} N_n = 3$
Total thermodynamic energy of system $E = \sum\limits_{n} N_n \varepsilon_n = 3\varepsilon$
Total number of macrostates = 3
Total number of microstates = $\sum\limits_{i} \omega_i = 8 + 6 + 4 = 18$

Consequently, our very simple thermodynamic system is describable by three possible macrostates and 18 associated microstates. Although this is a completely unreal example physically speaking (since we cannot think of ever creating such a high vacuum in a laboratory container) we conclude that if we had three particles in our container, as time passed and the particles exchanged translational energy amongst themselves by elastic collisions and also elastic collisions with the container walls, we might expect that all three macrostates and their associated 18 microstates would occur, disappear, then reappear and so on. If we assume that any particular microstate appears randomly and without preference, that all microstates have the same probability of occurring then we can define the (thermodynamic) probability of a microstate as,

Probability of a microstate = $\dfrac{1}{\text{Total number of microstates for all allowed macrostates of the thermodynamic system}}$

Likewise we can define the (thermodynamic) probability of a macrostate as,

Probability of a macrostate = $\dfrac{\text{Number of microstates associated with a particular macrostate}}{\text{Total number of microstates for all allowed macrostates of the thermodynamic system}}$

Hence

Probability of macrostate 1 (i.e. probability of macrostate 1 appearing)
$= \frac{8}{18} = 0.44$

Probability of macrostate 2 (i.e. probability of macrostate 2 appearing)
$= \frac{6}{18} = 0.33$

Probability of macrostate 3 (i.e. probability of macrostate 3 appearing)
$= \frac{4}{18} = 0.22$

Statistical Thermodynamics of the Thermally Perfect Gas

since each macrostate appears $\dfrac{\omega_i}{\sum_i \omega_i}$ of the time. We could conclude from this that the macrostate with the greatest number of associated microstates viz. microstate 1 will occur most frequently. Stated another way, there is a 'preferred' or 'most probable' macrostate of this simple system (macrostate 1) that will on average appear somewhat more often than either of the other two macrostates *purely because macrostate 1 has got the most microstates associated with it*. This idea of a preferred or most probable macrostate is very important because when we come to deal with realistic thermodynamic systems containing many millions of particles, the preferred or most probable macrostate has an overwhelmingly larger number of microstates associated with it than any other macrostate with its attendant microstates and in fact the preferred or most probable macrostate turns out to be identifiable with the thermodynamic state of the system.

Let us now leave this simple problem and proceed quite generally, by determining the macrostates and microstates of our thermodynamic system of N particles and look for some particular macrostate that is more likely to occur. In order to apply statistical methods we must ensure that we only apply statistics to large numbers. Let us, as we have already commented, 'collect' our particles into groups of translational energy levels where each and every particle in the group possesses essentially the same energy $\bar{\varepsilon}_n$. Let there be $\bar{N}_n$ particles in the group of energy $\bar{\varepsilon}_n$ and let the total statistical weight of the group be $\bar{g}_n$. We need not decide how big $\bar{N}_n$ need be since we shall later drop the grouping feature. We only use the grouping concept so that we can apply statistics and abandon the groups when the statistical analysis is complete, since they are not physically significant. We can, however, get some idea of the numbers involved. We have seen in Section 2.4 that the number of translational states accessible to an 'average' helium atom is about 10^{28}. Hence, even if we divide up these energy states into, say, 10^8 different groups of energy states then each group will still contain about 10^{20} energy states. Furthermore, since our helium atom has a translational energy of about 10^{-21} J whilst the energy separation between successive translational energy levels is about 10^{-48} J we see that a group of 10^{20} levels could represent a maximum energy spread of about 10^{-28} J, or about 10^{-7} of the energy of the atom. Hence we can rightly say that the energy of each level in the group is essentially $\bar{\varepsilon}_n$. Figure 2.5 shows a diagrammatic representation of the grouping of translational energy levels.

2.6 The Pauli exclusion principle

So far we have made no restriction on the number of particles allowed to occupy any 'part-rung' or energy state that forms part of any 'rung' or degenerate energy level. For example, consider macrostate 2 for which we have shown that microstate 1 of this macrostate is characterized by

$$N_0 = 2, 0; N_1 = 0, 0; N_2 = 0, 0; N_3 = 1, 0$$

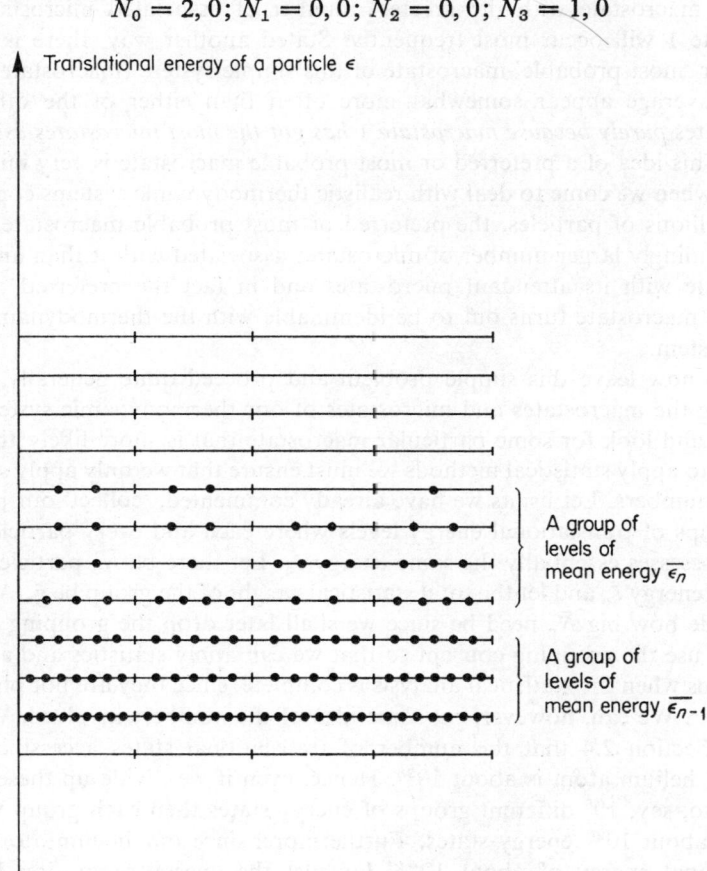

Figure 2.5. A schematic representation of the grouping of translational energy levels into groups characterized by a common energy. In the figure the group of levels characterized by mean energy $\bar{\varepsilon}_n$ comprises 16 energy states and 36 identical particles. The group of levels characterized by mean energy $\bar{\varepsilon}_{n-1}$ comprises 12 energy states and 76 identical particles.

This assumes that it is permitted to put more than one particle on the 'part-rung' energy state forming part of the 'rung' or translational energy level ε_0 (in case the reader has forgotten we remind him that the 'part-rung' is an energy state 'within' a 'rung' or an energy level!).

This is a very important basic point; either there is a restriction on the number

Statistical Thermodynamics of the Thermally Perfect Gas

of particles that can occupy any energy state or there is no restriction. We shall distinguish between two classes of particles termed 'bosons' and 'fermions'. Bosons are particles unrestricted in number in their allocation to translational energy states, whereas fermions are restricted. The reason why one class of particles is unrestricted whilst the other class is restricted involves quantum theory and we shall not discuss it in a text on thermodynamics; we shall merely accept it. The fact that there are two classes of particles gives rise to two types of problem to analyse statistically when we count our macrostates and microstates. Bose–Einstein (B.E.) statistics deals with bosons (the 'all-in together' particles) whilst Fermi–Dirac (F.D.) statistics deals with fermions (the 'very exclusive' particles). In fact, the mathematical, statistical methods that we shall use are the *same* for both B.E. and F.D. statistics, the only difference being whether we begin this statistical analysis by allowing for any restrictions on the numbers of particles allowed to occupy a translational energy state. It is also sensible to add, at this stage, that there is another class of particles called Maxwell–Boltzmann (M.B.) particles. The reader should appreciate that M.B. statistics is not really a 'distinct' form of statistics; rather it is the common limiting form of *both* B.E. and F.D. statistics, as we shall show later. We shall discuss M.B. statistics in detail in Chapter 3 and state here that we do not need to ask the question as to whether there is any restriction on the number of particles allowed to occupy a translational energy state. The reason is simply that there are so many translational energy states available and so few particles to put into them that the question of there being more than one particle in a translational energy state does not arise—many, many energy states are unoccupied (in fact most of them) and the remaining quantum states are 'lucky' to have a particle at all! We shall return to this idea of a particle number restriction (known as the Pauli exclusion principle) when we discuss F.D. statistics; it does not apply to B.E. statistics.

2.7 Bose–Einstein statistics

Here the particles do not obey the Pauli principle, that is, they are not restricted in any way in the manner in which they are distributed amongst the various translational energy states. We are going to determine the number of macrostates with their associated microstates that are allowed for the problem of a thermodynamic system consisting of N indistinguishable, independent non-interacting particles moving in a container of volume V such that the particles are all bosons and also the system is in a known constant, static thermodynamic state and such that

$$\sum_n N_n = N$$
$$\sum_n \varepsilon_n N_n = E$$

Let us consider that group of levels in Figure 2.5 which is characterized by a mean translational energy $\bar{\varepsilon}_n$, a total number of particles $\bar{N}_n$ and a total statistical weight of $\bar{g}_n$. Our problem, which is entirely analogous to Figure 2.4 and Table 2.2, is to find out how many ways there are (t_n) that we can put the $\bar{N}_n$ identical particles into the $\bar{g}_n$ translational energy states. Since we have pictured the translational energy states as being 'part-rungs' let us imagine the following 'thought' experiment. We lay all the $\bar{g}_n$ 'part-rungs' in a straight line and assume that the partitions (but *not* the two end pieces) that separate each 'part-rung'

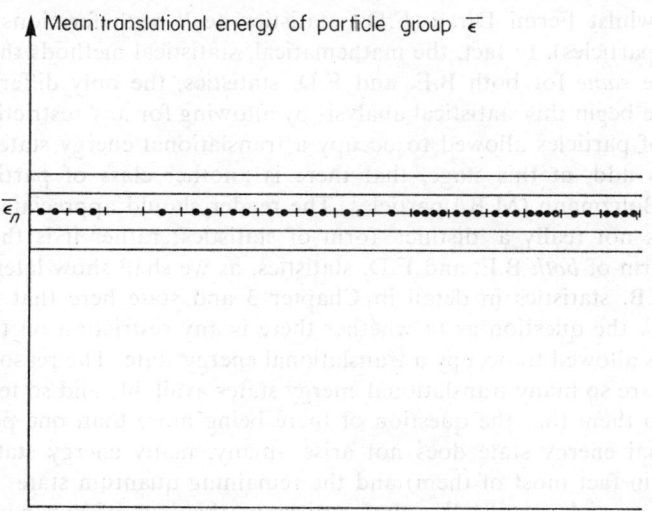

Figure 2.6. The result of grouping the translational energy levels $\bar{\varepsilon}_n$ of Figure 2.5 into one energy level group. This single level group comprises 16 energy states all of the same energy and has a statistical weight of 16. The occupancy of this level group is 36 identical particles.

from its neighbour are movable. Since there are $\bar{g}_n$ 'part-rungs' there are $(\bar{g}_n - 1)$ movable partitions.

Figure 2.6 shows the complete 'rung' (which represents all the energy levels of the group), the 'part-rungs' of these energy levels (where each 'part-rung' represents an energy state) together with the movable partitions for the case $\bar{g}_n = 16$.

We must now fill the 'part-rungs' with our $\bar{N}_n$ particles. Figure 2.6 shows a 'rung' consisting of 16 'part-rungs', 15 movable partitions containing a total of 36 particles. We note that there are two particles on the first 'part-rung', two on the second, none on the third, etc. To determine the number of different ways of putting these $\bar{N}_n$ particles on the $\bar{g}_n$ 'part-rungs', let the particles and

Statistical Thermodynamics of the Thermally Perfect Gas

the movable partitions be considered as a collection of $(\bar{N}_n + \bar{g}_n - 1)$ objects lying in a straight line to be arranged amongst each other. If all the particles and all the movable partitions were distinguishable from each other then the number of arrangements would be $(\bar{N}_n + \bar{g}_n - 1)!$. For the example of Figure 2.6 this would be $(36 + 16 - 1)!$ or $51!$. However, the particles are identical and therefore not distinguishable from each other so that rearranging the particles amongst themselves would not lead to different arrangements. For example, in Figure 2.6 if we exchange the two particles on 'part-rung' 1 with the two particles on 'part-rung' 9 then things would still look the same; or if we exchanged any two of the particles on 'part-rung' 11 with any two on 'part-rung' 13 then things would also still look the same. We therefore must divide our total number of arrangements just deduced by the number of arrangements (permutations) amongst the $\bar{N}_n$ particles and get

$$\frac{(\bar{N}_n + \bar{g}_n - 1)!}{\bar{N}_n!}$$

However, this is not all, because as well as all of the particles being identical so are the $(\bar{g}_n - 1)$ movable partitions so that we can exchange these identical partitions amongst themselves and things will still look the same. We must therefore divide the above number of arrangements by the number of arrangements (permutations) of the $(\bar{g}_n - 1)$ identical things amongst themselves. We finally obtain

Number of distinct and different ways of allocating $\bar{N}_n$ particles amongst $\bar{g}_n$ translational energy states with no restriction on particle numbers in any translational energy state

$$= t_n$$

$$= \frac{(\bar{N}_n + \bar{g}_n - 1)!}{\bar{N}_n!\,(\bar{g}_n - 1)!} \tag{2.16}$$

The quantity t_n has been solely concerned with that particular group of levels of energy $\bar{\varepsilon}_n$ in Figure 2.5 characterized by $\bar{N}_n$ particles and $\bar{g}_n$ translational energy states, and it takes no account of any of the other groups. For example for the group containing $\bar{N}_{n-1}$ particles, $\bar{g}_{n-1}$ translational energy states all characterized by energy $\bar{\varepsilon}_{n-1}$ we have, by reasoning similar to the above,

$$t_{n-1} = \frac{(\bar{N}_{n-1} + \bar{g}_{n-1} - 1)!}{(\bar{N}_{n-1})!\,(\bar{g}_{n-1} - 1)!} \tag{2.17}$$

If then we consider the group of energy $\bar{\varepsilon}_n$ and the group of energy $\bar{\varepsilon}_{n-1}$ together we can say that since any one distinct arrangement of the former can always be linked with *any* one distinct arrangement of the latter then

Number of distinct and different ways of allocating $\bar{N}_n$ particles into $\bar{g}_n$ translational energy states and $\bar{N}_{n-1}$ particles into $\bar{g}_{n-1}$ translational energy states
$$= t_n \cdot t_{n-1}$$

Clearly this can be extended to consider arrangements for each and every group of levels taken altogether. If therefore we consider all of the above N particles distributed throughout all of the various grouped levels we can say that

Number of distinct and different ways of arranging

$\bar{N}_0$ particles in group of energy $\bar{\varepsilon}_0$,
$\bar{N}_1$ particles in group of energy $\bar{\varepsilon}_1$, $= \omega_{\bar{N}_0, \bar{N}_1, \bar{N}_2, \ldots \bar{N}_n, \ldots}$
$\bar{N}_2$ particles in group of energy $\bar{\varepsilon}_2$,
etc.
$\bar{N}_n$ particles in group of energy $\bar{\varepsilon}_n$,
etc.
$$= \prod_n t_n$$
$$= \prod_n \frac{(\bar{N}_n + \bar{g}_n - 1)!}{\bar{N}_n!(g_n - 1)!} \quad (2.18)$$

This is, in fact, the total number of microstates corresponding to the macrostate described on the left-hand side of the equation (2.18).

But we saw in our simple example in Section 2.5 that there are other macrostates which also are allowed, since these macrostates also satisfy equations (2.2) and (2.3) (which can be written for the present case where we have grouped levels for statistical purposes)

$$\Sigma \bar{N}_n = N \quad (2.19)$$

and

$$\Sigma \bar{N}_n \bar{\varepsilon}_n = E \quad (2.20)$$

Hence

Total number of microstates corresponding to all macrostates that satisfy equations (2.19) and (2.20)
$$= \Omega_{\text{BE}}$$
$$= \underset{\substack{\text{all} \\ \text{allowed} \\ \text{macrostates}}}{\Sigma} \omega_{\bar{N}_0, \bar{N}_1, \ldots \bar{N}_n, \ldots}$$

Statistical Thermodynamics of the Thermally Perfect Gas

i.e.

$$\Omega_{BE} = \sum_{\substack{\text{all} \\ \text{allowed} \\ \text{macrostates}}} \prod_n \frac{(\bar{N}_n + \bar{g}_n - 1)!}{\bar{N}_n!(\bar{g}_n - 1)!} \qquad (2.21)$$

Here we have put the suffix BE to show that this result only holds for bosons which are unrestricted in the sense of Section 2.6. Equation (2.21) represents the total number of microstates associated with all macrostates that can occur for our thermodynamic system of bosons and all of these microstates are assumed to be equally probable, that is equally likely to occur. Let us check that the simple example considered in Section 2.5 agrees with this formula. There we shows that there were three macrostates, the first one of which was

Macrostate 1 $N_0 = 1; N_1 = 1; N_2 = 1; N_3 = 0$ etc.; $g = 2$ for all states.

Total number of microstates associated with this macrostate

$$= \prod_n \frac{(\bar{N}_n + \bar{g}_n - 1)!}{\bar{N}_n!(\bar{g}_n - 1)!}$$

$$= \prod_3 \frac{(1 + 2 - 1)!}{1!(2 - 1)!}$$

$$= 2 . 2 . 2 = 8$$

and these are the eight microstates shown in Figure 2.4.
Likewise, we have

Macrostate 2 $N_0 = 2; N_1 = 0; N_2 = 0; N_3 = 1; N_4 = 0$ etc.; $g = 2$ for all states

Total number of microstates associated with this macrostate

$$= \frac{(2 + 2 - 1)!}{2!1!} \frac{(0 + 2 - 1)!}{0!1!} \frac{(0 + 2 - 1)!}{0!1!} \frac{(1 + 2 - 1)!}{1!1!}$$

$$= 3 \times 1 \times 1 \times 2 \text{ since } 0! \equiv 1$$

$$= 6$$

and these are the microstates listed in Table 2.2.
Finally

Macrostate 3 $N_0 = 0; N_1 = 3; N_2 = 0; N_3 = 0$

Total number of microstates associated with this macrostate

$$= \frac{(0 + 2 - 1)!}{0!1!} \frac{(3 + 2 - 1)!}{3!1!} \frac{(0 + 2 - 1)!}{0!1!}$$

$$= 4$$

and these are the four microstates listed in Table 2.3. Hence,

$$\omega_{1,1,1,0,0...} = 8$$
$$\omega_{2,0,0,1,0...} = 6$$
$$\omega_{0,3,0,0,...} = 4$$

and

$$\Omega_{\mathrm{BE}} = \sum_i \omega_i = 8 + 6 + 4 = 18$$

Hence our formulae agree with the numbers deduced by trial and error in Section 2.5. We shall return to this mathematical series for Ω_{BE} in equation (2.21) when we have obtained the comparable expression in F.D. statistics.

2.8 Fermi–Dirac statistics

In F.D. statistics the Pauli exclusion principle states that there can never be more than one particle in any single-particle translational energy state. The Pauli principle applies to electrons but not to 'ordinary' gas molecules or atoms or phonons. We shall defer until Chapter 4 the exact meaning of the words 'single-particle translational energy state' of an electron; let us, for the time being identify such a state with a 'part-rung'. As we remarked earlier, the reasons for this 'exclusiveness' of electrons lies deeply within quantum theory; it is ultimately related to the mutual repulsion of electron wave functions but need not be considered in a book on thermodynamics.

Before making a general analysis of the microstates and macrostates of a gas of fermions let us consider the effect that the Pauli principle would have on the simple problem considered in Section 2.5. Here we considered a container holding only three identical particles such that the degeneracy of all of the energy levels was twofold and the total translational energy of the thermodynamic system was 3ε. The analysis in Section 2.5 showed that:

Macrostate 1.

There were eight associated microstates but none of these eight microstates had more than one particle on a 'part-rung' or single-particle translational energy state.

Macrostate 2.

There were six associated microstates listed in Table 2.2 Clearly if we impose the Pauli principle then microstates 1, 2, 3, 4 must be forbidden since they each involve two particles being on a 'part-rung' (i.e. single-particle translational state). However, microstates 5 and 6 are allowed since neither of them involves multiple occupancy of a 'part-rung', single-particle translational energy state.

Statistical Thermodynamics of the Thermally Perfect Gas

Macrostate 3.

If we apply the exclusion principle to Table 2.3 then we find that *all* of these microstates are forbidden since they all involve more than one particle being on a 'part-rung' or single-particle translational energy state.

Hence, if the particles in the problem of Section 2.5 had been fermions instead of bosons, we would have concluded that

Total number of macrostates = 2 (compared with 3 for B.E.)

Total number of microstates = 8 + 2 = 10 (compared with 18 for B.E.)

We see that the Pauli principle reduces the number of macrostates and associated microstates.

Let us now proceed to a general analysis where as in Section 2.7 we shall once again group our array of translational energy levels such that there are $\bar{N}_n$ particles in a group characterized by an average translational energy $\bar{\varepsilon}_n$ and a statistical weight $\bar{g}_n$. Because of the Pauli principle there can never be more than one particle in each single-particle translational energy state, that is, $\bar{N}_n$ can never be greater than $\bar{g}_n$. Put colloquially, there must always be more quantum states than particles to occupy them. Let us now consider that group of levels characterized by energy $\bar{\varepsilon}_n$. The number of different ways of placing the $\bar{N}_n$ particles on the $\bar{g}_n$ 'part-rungs' is found by supposing initially that the particles are distinguishable. We could then pick the 'first' particle and put it onto any one of the 'part-rungs'—there are $\bar{g}_n$ ways of doing this and for each and every one of these choices the 'second' particle could be put on any of the remaining $(\bar{g}_n - 1)$ 'part-rungs' and so on. Hence, if the particles were distinguishable we could say that the total number of different ways of placing the $\bar{N}_n$ particles on the $\bar{g}_n$ 'part-rungs' would be

$$\bar{g}_n(\bar{g}_n - 1)(\bar{g}_n - 2)\ldots(\bar{g}_n - \{\bar{N}_n - 1\}) = \frac{\bar{g}_n!}{(\bar{g}_n - \bar{N}_n)!}$$

However, the particles are not distinguishable—they are identical. This means that we could exchange any particle on any 'part-rung' with any other particle on any other 'part-rung' and things would still look the same, that is, we could permute the $\bar{N}_n$ particles amongst themselves and still leave everything unchanged. Since this can be done $\bar{N}_n!$ ways we conclude analogously with equation (2.16) for bosons

Number of distinct and different ways of allocating $\bar{N}_n$ particles amongst $\bar{g}_n$ single-particle translational energy states allowing for the exclusion principle restricting the number of particles to one per single-particle translational energy state

$$= t_n$$
$$= \frac{\bar{g}_n!}{(\bar{g}_n - \bar{N}_n)!\bar{N}_n!} \qquad (2.22)$$

Figure 2.7 shows the 'rungs', the 'part-rungs' and a possible particle allocation for the case of seven particles and 16 single particle translational energy states.

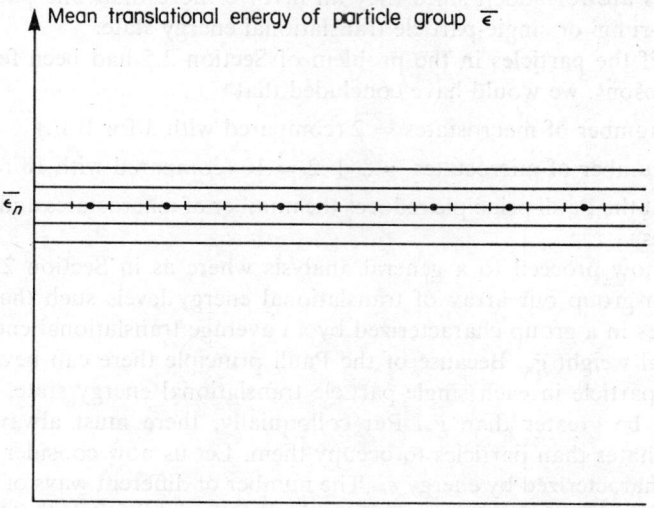

Figure 2.7. The result of grouping a set of energy levels into a single energy level group of mean energy $\bar{\varepsilon}_n$. This single level group comprises 16 single particle translational energy states all of the same energy and has a statistical weight of 16. The occupancy of the level is seven identical particles. This grouping is compatible with Fermi–Dirac statistics, since there is never any more than one particle in each single-particle translational energy state.

Following reasoning identical to Section 2.7 we can conclude

Total number of microstates corresponding to all macrostates that satisfy equations (2.19) and (2.20) and also the Pauli principle

$$= \Omega_{\text{FD}}$$
$$= \sum_{\substack{\text{all} \\ \text{allowed} \\ \text{macrostates}}} \omega_{N_0, N_1, N_2, \ldots N_n, \ldots}$$

i.e.

$$\Omega_{\text{FD}} = \sum_{\substack{\text{all} \\ \text{allowed} \\ \text{macrostates}}} \prod_n \frac{\bar{g}_n!}{(\bar{g}_n - \bar{N}_n)! \bar{N}_n!} \tag{2.23}$$

This represents the total number of microstates associated with all macrostates that can occur for our thermodynamic system of fermions and all these microstates are assumed to be equally probable, thus equally likely to occur.

Statistical Thermodynamics of the Thermally Perfect Gas

We can likewise show that the formulae above agree with the simple numerical example quoted at the beginning of this section; we leave this as an exercise for the reader.

2.9 Maxwell–Boltzmann statistics

In Section 2.6 we commented that there was a third class of statistics known as M.B. statistics but we also said that this class was not a distinct class of its own: rather, it is the common limiting form of *both* B.E. and F.D. statistics. In our discussion of F.D. statistics in Section 2.8 we found that we required $\tilde{g}_n \geqslant \bar{N}_n$ always—although such a restriction was not necessary in B.E. statistics since multiple translational energy-state occupancy was allowed.

Let us assume that for *both* F.D. and B.E. statistics

$$\bar{N}_n \ll \tilde{g}_n \qquad (2.24)$$

Consider B.E. statistics. Equation (2.16) is

$$t_n = \frac{(\bar{N}_n + \tilde{g}_n - 1)!}{\bar{N}_n!(\tilde{g}_n - 1)!}$$

$$= \frac{1}{\bar{N}_n!}[(\tilde{g}_n - 1) + \bar{N}_n] \cdot [(\tilde{g}_n - 1) + (\bar{N}_n - 1)] \ldots$$

$$[(\tilde{g}_n - 1) + \bar{N}_n - (\bar{N}_n - 1)]$$

multiplied by

$$\frac{(\tilde{g}_n - 1) + (\bar{N}_n - \bar{N}_n)}{(\tilde{g}_n - 1)} \cdot \frac{(\tilde{g}_n - 1) + \bar{N}_n - (\bar{N}_n + 1)}{(\tilde{g}_n - 1) - 1} \cdots$$

Making the approximation of equation (2.24) gives

$$t_n \sim \frac{1}{\bar{N}_n!}[\tilde{g}_n + (\bar{N}_n - 1)][\tilde{g}_n + (\bar{N}_n - 2)] \ldots [\tilde{g}_n]$$

$$= \frac{(\tilde{g}_n)^{\bar{N}_n}}{\bar{N}_n!} \qquad (2.25)$$

Hence analogously to equation (2.21) we have, in this limiting case

$$\Omega_{\text{MB}} = \sum_{\substack{\text{all} \\ \text{allowed} \\ \text{macrostates}}} \prod_n \frac{(\tilde{g}_n)^{\bar{N}_n}}{\bar{N}_n!} \qquad (2.26)$$

42   Thermodynamics of Electrical Processes

Consider now F.D. statistics. Equation (2.22) gives

$$t_n = \frac{\tilde{g}_n!}{(\tilde{g}_n - \bar{N}_n)!\bar{N}_n!}$$

$$= \frac{[\tilde{g}_n - 0][\tilde{g}_n - 1]\ldots[\tilde{g}_n - (\bar{N}_n - 1)]}{(\tilde{g}_n - \bar{N}_n)!\bar{N}_n!} \cdot \frac{[\tilde{g}_n - \bar{N}_n][\tilde{g}_n - (\bar{N}_n + 1)]}{1}\ldots$$

$$\sim \frac{[\tilde{g}_n - 0][\tilde{g}_n - 1]\ldots[\tilde{g}_n - (\bar{N}_n - 1)]}{\bar{N}_n!}$$

using the approximation of equation (2.24). Hence in this limiting case

$$t_n = \frac{(\tilde{g}_n)^{\bar{N}_n}}{\bar{N}_n!}$$

thereby giving us a result identical to equation (2.25) and also to equation (2.26) in turn. We conclude that the asymptotic form of *both* B.E. and F.D. statistics is the common form known as M.B. statistics according to equation (2.26).

We shall discuss the applications of M.B. statistics in Chapter 3. We note that whereas we have characterized B.E. particles as the 'all in together' particles, F.D. particles as 'very exclusive' particles we can characterize M.B. particles as 'few in number' particles.

2.10 The 'preferred' or 'most probable' macrostates

We now have three expressions for Ω given by equations (2.21), (2.23) and (2.26). In all three cases Ω is a mathematical series (as shown by the summation sign) whilst each term of the series is itself rather complicated, being a continued product of the form shown in equations (2.18) or (2.23). A detailed analysis of these series is an interesting application of the asymptotic behaviour of finite-series mathematics—but of little or no interest to the engineering or physics undergraduate! We shall not attempt a mathematical discussion at all. What we shall do is make a very dramatic assumption—we shall assume that within this series of very many terms (millions of them!) there is *one* term that is so large that *all* other terms of the series can be ignored, thus we will replace the entire summation of terms by just one single term. Obviously this is a very drastic step and we shall not attempt to justify this so-called 'maximum-term' method mathematically. We can, however, make it plausible by considering the very simple example discussed in Section 2.5 where we showed that that macrostate with the greatest number of associated microstates would appear somewhat more often than any of the other two allowed macrostates. If we had used realistic numbers, such as billions of particles and billions of translational

Statistical Thermodynamics of the Thermally Perfect Gas

energy states, then this conclusion would have come out very much more positively and dramatically; that is, there would be one macrostate (known as the preferred or most probable macrostate) which had so many microstates associated with it that no other macrostate would have 'had a chance' of being observed a significant number of times.

It is possible that the reader may now be confused by what appears to be a contradiction. Let us repeat what has been concluded. We initially assumed that subject to equations (2.2) and (2.3) [or equations (2.19) and (2.20)] our thermodynamic system in its fixed unchanging state of thermodynamic equilibrium would appear when 'viewed' microscopically in the form of allowed macrostates (of which there would be many) where each one of these many macrostates would have an even larger number of microstates associated with it. We also assumed (and always will assume) that all microstates are equally probable, that is, the system would *not* 'pick' a preferred or most probable microstate—each and every microstate would 'have its fair turn' of appearing.

However, we are now making the additional statement that there is one particular macrostate known as the preferred or most probable macrostate that has got such an enormous number of microstates associated with it (a number which is considerably larger than the number of microstates associated with any of the other non-preferred but allowed macrostates). Suppose, referring back to the example of Section 2.5, we had found that for realistic numbers of particles and energy states, there had been 10^{30} macrostates and that all but one of these macrostates had each about 10^{50} microstates associated with it but one macrostate (the preferred or most probable macrostate) had 10^{100} microstates associated with it. Then we could say

Total number of microstates for all macrostates

$$= (10^{30} - 1)(10^{50}) + 1(10^{100})$$
$$= 10^{80} + 10^{100}$$

Probability of each microstate

$$= \frac{1}{10^{80} + 10^{100}}$$

Probability of each and every unpreferred macrostate

$$= \frac{10^{50}}{10^{80} + 10^{100}} \sim 10^{-50}$$

Probability of the preferred macrostate

$$= \frac{10^{100}}{10^{80} + 10^{100}} \sim 1$$

Hence although the system would pass through *all* of the microstates, including all those of the very many unpreferred macrostates, then as time passed by because there are so very many microstates associated with the preferred or most probable macrostate, every time we had a 'look' at the system it would always overwhelmingly appear to be in one of the microstates associated with the preferred or most probable macrostate. Put in another way, the system spends the vast bulk of its time in passing through those microstates that belong to the preferred or most probable macrostate. However, we must remember that all the other microstates and macrostates are not excluded. They do in fact appear; but since they are so 'few in number' (relatively speaking!) they can be ignored. The reader should note that although there is a preferred or most probable macrostate there is *not* a preferred or most probable microstate—any microstate of any macrostate can turn up with an equal chance. Therefore, as each term in our series for Ω (in all three cases) represents the number of microstates associated with each allowed macrostate of the system, we can drop all of the terms except that one corresponding to the preferred or most probable macrostate. Accepting this very great assumption, equations (2.21), (2.23), (2.26) become, when we drop the summation sign and replace the entire series by a single term

$$\Omega \simeq (\Pi_n t_n)_{\text{preferred or most probable macrostate}} = \Pi_n t_n^* \qquad (2.27)$$

where the superscript 'star' (asterisk) refers to the preferred or most probable macrostate. In any realistic description of a macrostate the number $\bar{N}_n^*$ will be very large as will be the $\bar{g}_n$. For these reasons we can drop the unity term in equations (2.16), (2.18), (2.21) for F.D. statistics. Furthermore, since the quantity Ω is going to be very large it will be more convenient to discuss the properties of $\ln \Omega$ rather than Ω itself. There is no loss in generality in doing this and it enables up to simplify the analysis because we can replace the continued product operator by a simple summation. Hence,

$$\left. \begin{array}{l} \ln \Omega_{\text{BE}} = \sum_n [\ln (\bar{g}_n - \bar{N}_n^*)! - \ln \bar{g}_n! - \ln \bar{N}_n^*!] \\ \ln \Omega_{\text{FD}} = \sum_n [\ln \bar{g}_n! - \ln (\bar{g}_n - \bar{N}_n^*)! - \ln \bar{N}_n^*!] \\ \ln \Omega_{\text{MB}} = \sum_n [\ln (\bar{g}_n)^{\bar{N}_n^*} - \ln \bar{N}_n^*!] \end{array} \right\} \qquad (2.28)$$

These can be simplified even further since the logarithm of the factorial of a large number can be expanded by Stirling's theorem, viz.

$$\ln x! \simeq x \ln (x) - x$$

Statistical Thermodynamics of the Thermally Perfect Gas

provided x is greater than 10, say. If we use Stirling's theorem and perform a little algebra we get

$$\left. \begin{array}{l} \ln \Omega_{BE} = \sum_n \left[\tilde{g}_n \ln \left(1 + \frac{\bar{N}_n^*}{\tilde{g}_n} \right) + \bar{N}_n^* \ln \left(\frac{\tilde{g}_n}{\bar{N}_n^*} + 1 \right) \right] \\[2mm] \ln \Omega_{FD} = \sum_n \left[-\tilde{g}_n \ln \left(1 - \frac{\bar{N}_n^*}{\tilde{g}_n} \right) + \bar{N}_n^* \ln \left(\frac{\tilde{g}_n}{\bar{N}_n^*} - 1 \right) \right] \\[2mm] \ln \Omega_{MB} = \sum_n \left[0 + \bar{N}_n^* \ln \left(\frac{\tilde{g}_n}{\bar{N}_n^*} + 1 \right) \right] \end{array} \right\} \quad (2.29)$$

We have already said several times that the preferred or most probable macrostate is that one which has the greatest number of associated microstates, and we used this assumption when we approximated our series for Ω by just a single term. However, we can frame this 'greatest number of microstates' assumption mathematically to tell us something about the set of numbers

$$\bar{N}_0^*, \bar{N}_1^*, \ldots \bar{N}_n^* \ldots$$

that actually represent this preferred or most probable macrostate. We cannot choose this set of numbers in any way we like since they must obey equations (2.19) and (2.20). For example there would be no point in choosing $\bar{N}_2^* = 10^{24}$ particles (i.e. 10^{24} particles in the second group of energy levels) if there were only 10^{22} particles in the entire thermodynamic system. Likewise it would be wrong to have the total translational energy in any group of levels greater than the total thermodynamic energy E of the entire thermodynamic system.

We must therefore look for that set of number $\bar{N}_n^*$ that gives the maximum term in Ω and yet obeys equations (2.19, and 2.20). To do this we use Lagrange's method of undetermined multipliers which is no more than a mathematical procedure enabling us to find the maximum or minimum of a function of many variables subject to certain restrictive conditions. In this case the function is $\ln \Omega$ and the variables are $\bar{N}_0^*, \bar{N}_1^*, \ldots \bar{N}_n^*, \ldots$. These variables are *not* all independent because they are subject to the restrictive conditions of equations (2.19) and (2.20).

To obtain the maximum of our function we consider a differential increment of the function.

$$\mathrm{d} \ln \Omega = \sum_n \frac{\partial}{\partial \bar{N}_n^*} (\ln \Omega_{BE/FD})$$

$$= \sum_n \left[\pm \frac{\tilde{g}_n \left(\pm \frac{1}{\tilde{g}_n} \right)}{1 \pm \frac{\bar{N}_n^*}{\tilde{g}_n}} + \ln \left(\frac{\tilde{g}_n}{\bar{N}_n^*} \pm 1 \right) - \frac{\frac{\tilde{g}_n}{\bar{N}_n^*}}{\left(\frac{\tilde{g}_n}{\bar{N}_n^*} \pm 1 \right)} \right] \mathrm{d} \bar{N}_n^*$$

All the plus signs are B.E. and all the minus signs are F.D., i.e.

$$\left.\begin{array}{l} d \ln \Omega_{\text{BE/FD}} = \sum_n \ln \left(\dfrac{\bar{g}_n}{\bar{N}_n^*} \pm 1 \right) d\bar{N}_n^* \text{ for B.E. and F.D. particles} \\ \text{and} \\ d \ln \Omega_{\text{MB}} = \sum_n \ln \left(\dfrac{\bar{g}_n}{\bar{N}_n^*} \right) d\bar{N}_n^* \text{ for M.B. particles} \end{array}\right\} \quad (2.30)$$

Normally, to find the maximum or minimum of a function we would now put the increment of this function equal to zero, i.e.

$$d \ln \Omega = 0 \quad (2.31)$$

This would be correct if all the increments of the variables were independent and *if* this were the case the only solution would be

$$\ln \left[\frac{\bar{g}_n}{\bar{N}_n^*} \pm \left(\text{or} \begin{array}{c} 1 \\ 0 \end{array} \right) \right] = 0 \quad (2.32)$$

Here we have used the shorthand notation $\pm \left(\text{or} \begin{array}{c} 1 \\ 0 \end{array} \right)$ to indicate that the second term on the left-hand side of equation (2.32) can either be $+1$(B.E.), -1(F.D.), or 0(M.B.).

However, the increments $d\bar{N}_n^*$ in the variables $\bar{N}_n^*$ are not all independent of each other because the two equations of constraint in equations (2.19) and (2.20) restrict our freedom to vary the $d\bar{N}_n^*$ independently of each other. The physical significance of this is that since the total number of particles is constant then any increases in the populations of some energy-level groups must just be balanced by decreases in the populations of other energy-level groups. A similar conclusion holds for the energies of the groups of levels since the total energy is constant. In fact, only $(N-2)$ of the increments $d\bar{N}_n^*$ can be varied independently and once this is done the remaining two increments are fixed—see below. Let us suppose that, for example, the two increments $d\bar{N}_0^*$ and $d\bar{N}_1^*$ are fixed in this way. Then we have

$$d(\sum_n \bar{N}_n^*) = dN = 0 \quad (2.33)$$

since $N = $ constant. Also

$$d(\sum \bar{N}_n^* \cdot \bar{\varepsilon}_n) = dE = 0$$

i.e.

$$\sum_n \bar{\varepsilon}_n \cdot d\bar{N}_n^* = 0 \quad (2.34)$$

since $E = $ constant.

Statistical Thermodynamics of the Thermally Perfect Gas

We can solve equations (2.33) and (2.34) and find that

$$d\bar{N}_0^* = f_1(d\bar{N}_2^*, d\bar{N}_3^* \ldots)$$

and

$$d\bar{N}_1^* = f_2(d\bar{N}_2^*, d\bar{N}_3^* \ldots)$$

where f_1 and f_2 are mathematical functions. We can then substitute these values of $d\bar{N}_0^*$ and $d\bar{N}_1^*$ into equation (2.30), thereby eliminating $d\bar{N}_0^*$ and $d\bar{N}_1^*$ explicitly from equation (2.30), in which case we could then solve the resulting equation for the maximum. However, it is a rather messy algebriac procedure to find the two functions f_1 and f_2. It is more elegant, mathematically, to preserve the 'symmetry' of the algebraic equations by using Lagrange's method of undetermined multipliers to eliminate the dependent increments $d\bar{N}_0^*$, $d\bar{N}_1^*$.

The Lagrange method introduces two undetermined multipliers α and β such as to non-dimensionalize equations (2.33) and (2.34), thus we introduce α such that

$$\alpha \sum_n d\bar{N}_n^* = 0 \qquad (2.35)$$

where α has no dimensions since the $d\bar{N}_n^*$ likewise have no dimensions, being purely numbers. We also introduce β such that

$$\beta \sum_n \bar{\varepsilon}_n \cdot d\bar{N}_n^* = 0 \qquad (2.36)$$

Obviously β has the dimensions of reciprocal energy i.e. (energy)$^{-1}$ if equation (2.36) is to be dimensionless. Now that we have non-dimensionalized our two restrictive differential equations we can add equations (2.32), (2.35) and (2.36) to get

$$= \sum_n \left[\ln \left\{ \frac{\tilde{g}_n}{\bar{N}_n^*} \pm \begin{pmatrix} 1 \\ \text{or} \\ 0 \end{pmatrix} \right\} + \alpha + \beta \bar{\varepsilon}_n \right] d\bar{N}_n^* = 0 \qquad (2.37)$$

where the plus sign is for B.E., the minus sign for F.D. and the zero for M.B. We are at liberty to select the multipliers α and β so that the first two terms in the series on the right-hand side of equation (2.37) are always zero, i.e.

$$\ln \left\{ \frac{\tilde{g}_0}{\bar{N}_0^*} \pm \begin{pmatrix} 1 \\ \text{or} \\ 0 \end{pmatrix} \right\} + \alpha + \beta \bar{\varepsilon}_0 = 0$$

and

$$\ln \left\{ \frac{\tilde{g}_1}{\bar{N}_1^*} \pm \begin{pmatrix} 1 \\ \text{or} \\ 0 \end{pmatrix} \right\} + \alpha + \beta \bar{\varepsilon}_1 = 0$$

Equation (2.37) then becomes

$$= \sum_{k \geqslant 2} \left[\ln \left\{ \frac{\tilde{g}_k}{\bar{N}_k^*} \pm \begin{pmatrix} 1 \\ \text{or} \\ 0 \end{pmatrix} \right\} + \alpha + \beta \bar{\varepsilon}_k \right] d\bar{N}_k^* = 0 \qquad (2.37a)$$

Now in this latter equation, unlike equation (2.37), all of the increments $d\bar{N}_k^*$ are now *all* completely independent. The only possible solution to equation (2.37a) with the right-hand side equal to zero is that each and every term of the series on the left-hand side is identically zero, i.e.

$$\ln\left\{\frac{\bar{g}_k}{\bar{N}_k^*} \pm \begin{pmatrix}1\\ \text{or}\\ 0\end{pmatrix}\right\} + \alpha + \beta\bar{\varepsilon}_k = 0$$

Hence our general solution is that, one way or another, *all* of the coefficients of the increments $d\bar{N}_n^*$ in equation (2.37) are zero, i.e.

$$\ln\left\{\frac{\bar{g}_n}{\bar{N}_n^*} \pm \begin{pmatrix}1\\ \text{or}\\ 0\end{pmatrix}\right\} + \alpha + \beta\bar{\varepsilon}_n = 0$$

whence

$$\frac{\bar{N}_n^*}{\bar{g}_n} = \frac{1}{\exp(-\alpha) \cdot \exp(-\beta\bar{\varepsilon}_n) \mp \begin{pmatrix}1\\ \text{or}\\ 0\end{pmatrix}} \quad (2.38)$$

Here the *minus* sign is for B.E., the *plus* is for F.D. and the zero is for M.B.

Let us summarize, repeating earlier conclusions, just what this result means. We took a closed thermodynamic system of known E, V, T and analysed the energies of the N particles of the system as time passed by. We found that at any and every instant of time the system could be thought of as being in one of an array of macrostates and at that same time it would also be in one of an array of microstates. There were a vast number of microstates of the system and every such microstate belonged to one of the array of macrostates and had an equal chance of appearing. As time passes, the system will randomly pass through all of the microstates of each and every macrostate. However, since there was one macrostate (the preferred or most probable macrostate) that had so many associated microstates we would 'usually always' find the system in one of that vast number of microstates belonging to the preferred or most probable macrostate. We would therefore conclude that the gas will essentially always be found in the same macrostate as time passes although its microstate is always changing—usually, however, to just another microstate of the preferred or most probable macrostate.

We have now, in equation (2.38), found the formula actually giving the $\bar{N}_n^*$, the populations of the groups of levels of this preferred or most probable macrostate in terms of the $\bar{g}_n$ and $\bar{\varepsilon}_n$ which are known and in terms of α and β which are as yet undetermined although we do know their dimensions.

We can now drop the grouping of energy levels that was made in Sections 2.5, 2.7 and 2.8. The quantity on the left-hand side of equation (2.38) is known at the occupation index and represents the average number of particles (taken over a group of levels) found in any one energy level of that group of levels for

Statistical Thermodynamics of the Thermally Perfect Gas

the preferred or most probable macrostate. Clearly we can replace this by $\dfrac{N_n^*}{g_n}$ which is the average number of particles taken over just *one* level found in any one energy state of that level for the preferred or most probable macrostate.

In addition, we shall assume (and prove in Section 2.13) that

$$\beta = -\frac{1}{kT}$$

where k = Boltzmann's constant; T = absolute temperature of the system.

Our formulae for the preferred or most probable macrostate now become

$$\frac{N_n^*}{g_n} = \frac{1}{\exp(-\alpha)\exp\left(\dfrac{\varepsilon_n}{kT}\right) \mp \left(\substack{1\\ \text{or}\\ 0}\right)} \qquad (2.39)$$

where the plus sign is for F.D., the minus sign is for B.E., and the zero is for M.B. We see that *if* for both B.E. and F.D. statistics

$$\exp(-\alpha)\exp\left(\frac{\varepsilon_n}{kT}\right) \gg 1$$

then we can always ignore the unity term in the denominator and get the M.B. result. Therefore, one criterion for M.B. statistics to be a common limiting form of *both* B.E. and F.D. statistics is that

$$\exp(-\alpha)\exp\left(\frac{\varepsilon_n}{kT}\right) \gg 1 \qquad (2.40)$$

Figure 2.8 shows a plot of the occupation index for the three statistics as a function of $\left(\dfrac{\varepsilon_n}{kT} - \alpha\right)$. We note that:

a. the occupation index for F.D. statistics can never exceed unity. This is merely a restatement of the Pauli exclusion principle.

b. For a given value of α

$$\left(\frac{N_n^*}{g_n}\right)_{FD} < \left(\frac{N_n^*}{g_n}\right)_{MB} < \left(\frac{N_n^*}{g_n}\right)_{BE}$$

This means that B.E. statistics tend to concentrate the particles into the lower energies whilst F.D. statistics tend to concentrate them into higher energies when we compare them both with M.B. statistics. For this reason some authors conclude that this concentration in the B.E. case is the result of an 'effective' attraction between the particles whilst in the F.D. case it is a result of an 'effective' repulsion between the particles. However, as far as we are concerned our analyses have assumed thermally perfect gas behaviour, with independent particles having no mutual interactions. We have, of course, allowed indirectly for interactions by invoking the Pauli principle for fermions. We remind the reader of the distinction made at the beginning of Section 2.2 between mutual interaction (which is ignored) and collisional interaction which must be allowed to occur.

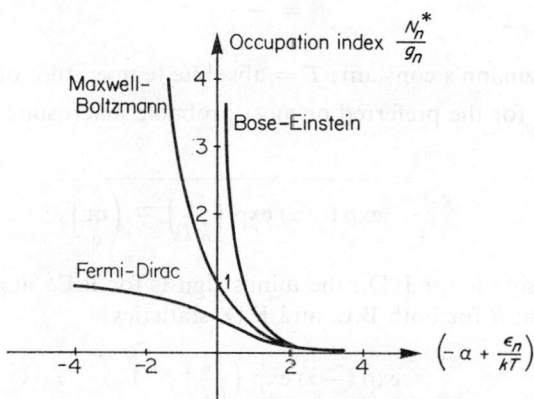

Figure 2.8. Occupation index (average number of particles in the nth translational energy state), as a function of dimensionless energy $-\alpha + \frac{\varepsilon_n}{kT}$ for Fermi–Dirac, Bose–Einstein and Maxwell–Boltzmann statistics. We note that for $-\alpha + \frac{\varepsilon_n}{kT} > 2$, say the three occupation indices are the same, meaning that there is no essential difference in the high energy state occupancy amongst the three statistics.

2.11 The links between a statistical and a thermodynamic description of the thermally perfect gas

We have now delved fairly deeply into the statistical aspects of a closed thermodynamic system. Let us see if we can relate these statistical aspects to the way we describe the system at the macroscopic (laboratory) level.

We have seen in Section 2.2 that the particles of our system are in incessant motion, constantly colliding amongst themselves and with the walls of their container and thereby constantly exchanging energy amongst themselves. Out of this picture of ceaseless activity we have constructed a pattern of behaviour such that every particle at each and every instant of time is in some quantized energy state and all of the particles could be described by a microstate. These microstates changed as time passed but they usually changed such that they all belonged to one macrostate—the preferred or most probable macrostate which

Statistical Thermodynamics of the Thermally Perfect Gas

does not significantly alter in relation to time. Clearly this preferred or most probable macrostate is identifiable with the fixed thermodynamic state of the system.

Thermodynamically speaking the system is described by equation (1.1)

$$đQ - đW = dE \qquad (2.41)$$

which is the First Law of Thermodynamics, and also equation (1.11)

$$TdS - pdV = dE \qquad (2.42)$$

which is the Second Law of Thermodynamics. Both of these equations hold for closed systems (systems whose total number of particles is constant) and we have in fact already used this when we wrote

$$\sum_n N_n^* = N$$

This is one link between a statistical microscopic description and a macroscopic (laboratory) description of the system since the left-hand side relates to quantized energy-state populations whilst the right-hand side is related to laboratory chemistry via the Avogadro hypothesis (see Appendix 2).

The second link is using equation (2.3) viz.

$$\sum_n N_n^* \varepsilon_n = E$$

To show the equivalence between a microscopic and a macroscopic thermodynamic description of the system let us suppose that the thermally perfect gas is in a container fitted with a piston and that it undergoes a change of thermodynamic state such that the energy of the system changes by dE. Such a change viewed macroscopically can only occur by a heat and/or a work transfer to the system since a mass transfer is disallowed in a closed system; this follows from the First Law (see equation 1.1). Let us further suppose that during this change of state the system is effectively in thermodynamic equilibrium at all times so that the change of state is thermodynamically reversible—see Section 1.5. Then, from equation (2.3) we have, on differentiation,

$$dE = \sum_n N_n^* d\varepsilon_n + \sum \varepsilon_n dN_n^* \qquad (2.43)$$

Let us consider the terms on the right-hand side of equation (2.43). The term $\Sigma N_n^* d\varepsilon_n$ results from a change in the value of the energy levels ε_n, thus they change their energies from being ε_n to being $\varepsilon_n + d\varepsilon_n$; however, the populations of particles in each energy state N_n^* are to remain unchanged. How can the energy-state populations remain fixed and yet the energy-state values change? This can only arise as a result of external forces acting on the system. We can prove this because for a thermally perfect gas in a container equation (2.9) shows that

$$\varepsilon_n = \frac{h^2}{8mV^{\frac{2}{3}}} \cdot (n_x^2 + n_y^2 + n_z^2)$$

i.e.

$$\varepsilon_n = C(n)V^{-\frac{2}{3}}$$

where $C(n)$ is a function of n but not of V.

Now

$$d\varepsilon_n = \left(\frac{\partial \varepsilon_n}{\partial V}\right)_n dV$$

$$\therefore \quad \sum_n N_n^* d\varepsilon_n = \sum_n N_n^* \left(\frac{\partial \varepsilon_n}{\partial V}\right) dV$$

$$= \left[\frac{\partial}{\partial V}(\Sigma N_n^* \varepsilon_n)\right]_{N_n^*} \cdot dV$$

$$= \left(\frac{\partial E}{\partial V}\right)_{N_n^*} \cdot dV$$

But from equation (2.42)

$$p = -\left(\frac{\partial E}{\partial V}\right)_{S, N_n^*}$$

Hence,

$$\Sigma N_n^* d\varepsilon_n = -p dV$$

or,

$$p = -\sum_n N_n^* \left(\frac{\partial \varepsilon_n}{\partial V}\right)_{N_n^*} \quad (2.44)$$

We can therefore identify $\sum_n N_n^* d\varepsilon_n$ with the mechanical work done *on* the system by the external force, (i.e. the work done is negative according to the sign convention of Chapter 1). Using equation (1.5) we can write

$$đW = -p dV$$
$$= \Sigma N_n^* d\varepsilon_n$$

There may be other work terms involved in changing the energy-state values but such changes will be in the values of potential energies (for particles possessing structure) and not in the translational kinetic energy values that we

Statistical Thermodynamics of the Thermally Perfect Gas

have considered here. For simplicity we will not consider such changes in potential energy but they can be important in special cases where electric and/or magnetic fields are applied to the system.

If we combine equations (2.43) and (2.44) we get

$$dE = \Sigma \varepsilon_n dN_n^* - đW \tag{2.45}$$

It appears reasonable to equate equations (2.45) and (2.41) to obtain

$$\Sigma \varepsilon_n dN_n^* = đQ \tag{2.46}$$

Let us consider the meaning of this equation which of course requires that the energy levels remain fixed in numerical value but all of the level populations change from N_n^* to $N_n^* + dN_n^*$. We take this to be a constant-volume reversible heat transfer $đQ_R$ to the system from the surroundings. During the change of state caused by this reversible heat transfer, the system is always in thermodynamic equilibrium with the surroundings, thus the system is always in the preferred or most probable macrostate so that Ω is a maximum.

In Section 2.10 we have shown that for all three statistics

$$d \ln \Omega + \alpha \sum_n dN_n^* + \beta \sum_n \varepsilon_n dN_n^* = 0 \tag{2.47}$$

where this equation follows from combining equations (2.30) and (2.37). Consider that particular case where no external forces act and therefore there is no displacement work transfer to or from the system; however, the system does undergo a reversible heat transfer with the surroundings. In this case equation (2.47) becomes

$$d \ln \Omega + \beta \Sigma \varepsilon_n dN_n^* = 0$$

i.e.
$$d \ln \Omega + \beta đQ_R = 0$$

using equation (2.46)

i.e.
$$d \ln \Omega - \frac{đQ_R}{kT} = 0$$

since we have assumed $\beta = -\dfrac{1}{kT}$

i.e.
$$d(k \ln \Omega) - \frac{đQ_R}{T} = 0$$

But for a reversible heat transfer under these restrictions we defined in thermodynamics the entropy function S such that [see equation (1.7)]

$$dS = \frac{dQ_R}{T} \tag{2.48}$$

Therefore

$$dS = d(k \ln \Omega)$$

whence

$$S = k \ln \Omega + \text{Constant} \tag{2.49}$$

This is another very important link between our microscopic and macroscopic concepts because the left-hand side describes the macroscopic property entropy whilst the right-hand side describes the logarithm of the total number of quantum microstates of the system.

If this had been a book on thermodynamics for mechanical engineers or physical chemists than equation (2.49) would have formed a starting point for a discussion of the Third Law of Thermodynamics [which gives the value of the constant on the right-hand side of equation (2.49)] together with a very clear physical picture of the meaning of both thermal and configurational entropy and its connexions with 'randomness and disorder'. However, limitations of space do not allow a complete discussion—see, nevertheless Section 8.2.

We can continue our comparison of the results of a quantum statistical analysis of the system comprising a thermally perfect gas by deducing the equation of state of a thermally perfect gas in all three statistics. Equation (2.44) shows that

$$p = -\sum_n N_n^* \left(\frac{\partial \varepsilon_n}{\partial V}\right)_{N_n^*}$$

We can interpret this equation by saying that the partial pressure exerted by a single particle in its nth translational energy state is

$$p_n = -\left(\frac{\partial \varepsilon_n}{\partial V}\right)_{N_n^*}$$

This is purely a mechanical relation and is concerned with the momentum change suffered by the particle as it collides with and rebounds elastically from the container walls. We can perform the differentiation because from equation (2.9) we have

$$\varepsilon_n = C(n) V^{-\frac{2}{3}}$$

$$\therefore \quad p_n = C(n) \cdot \tfrac{2}{3} V^{-\frac{5}{3}}$$

Statistical Thermodynamics of the Thermally Perfect Gas

Hence

$$\frac{p_n}{\varepsilon_n} = \frac{2}{3} \cdot \frac{C(n)V^{-\frac{5}{3}}}{C(n)V^{-\frac{2}{3}}}$$

$$= \tfrac{2}{3} V^{-1}$$

Now since the gas is thermally perfect it obeys the Gibbs–Dalton law

$$\sum_n p_n = p$$

i.e. the sum of partial pressures of all particles in all translational energy states is equal to the total pressure of the gas. Hence, by summation

$$\frac{p}{E} = \tfrac{2}{3} V^{-1}$$

i.e.

$$pV = \tfrac{2}{3} E \qquad (2.50)$$

This equation has been derived under purely mechanical considerations (momentum changes). It is the general equation of state of a thermally perfect gas made up of any class of particles provided that these particles have a non-zero rest mass (rest mass is essentially the mass for non-relativistic velocities). For that class of 'particles' having zero rest mass such as a photon or a phonon we shall show that the equation of state is

$$pV = \tfrac{1}{3} E \qquad (2.51)$$

2.12 The generalized partition function for B.E. and F.D. statistics

These is another important way of looking at the equivalence between a microscopic and a macroscopic description of a thermodynamic system. This involves introducing a mathematical function called the partition function. The partition function is no more than a mathematical series and it has no physical meaning. However it has great value in that it is a generating function, that is, the partial derivatives of the partition function give all the thermodynamic properties of the system. Let us see how we can introduce the partition function; we shall start with equation (2.39) for the preferred or most probable macrostate

$$N_n^* = \frac{g_n}{\exp(-\alpha)\exp\left(\dfrac{\varepsilon_n}{kT}\right) \pm 1}$$

where the plus sign is for F.D. and the minus sign is for B.E. Equations (2.2) and (2.3) then become

$$N = \sum_n N_n^* = \sum_n \frac{g_n}{\exp(-\alpha)\exp\left(\frac{\varepsilon_n}{kT}\right) \pm 1}$$

$$E = \sum_n N_n^* \varepsilon_n = \sum_n \frac{g_n \varepsilon_n}{\exp(-\alpha)\exp\left(\frac{\varepsilon_n}{kT}\right) \pm 1}$$

Let us introduce the partition function Z (it is only fair to state that this is being 'pulled out of the air' and is not being presented as a logical consequence of anything that has gone before!)

$$Z_{\text{BE/FD}} = \pm \sum_n g_n \ln\left[1 \pm \exp(\alpha)\exp\left(-\frac{\varepsilon_n}{kT}\right)\right] \qquad (2.52)$$

where the plus sign is still for F.D. and the minus sign for B.E.

Consider the derivative of equation (2.52)

$$\left(\frac{\partial Z_{\text{BE/FD}}}{\partial \alpha}\right)_{T,V} = \sum_n g_n \cdot \frac{1}{1 \pm \exp(\alpha)\exp\left(-\frac{\varepsilon_n}{kT}\right)} \cdot \exp(\alpha)\exp\left(-\frac{\varepsilon_n}{kT}\right)$$

where on the left-hand side we keep V constant in the differentiation by keeping the ε constant (recall that from equation (2.9), ε is a function of V) i.e.

$$\left(\frac{\partial Z_{\text{BE/FD}}}{\partial \alpha}\right)_{T,V} = \sum \frac{g_n}{\exp(-\alpha)\exp\left(\frac{\varepsilon_n}{kT}\right) \pm 1} = N \qquad (2.53)$$

This means that we have expressed N as a partial derivative of the partition function. By similar considerations involving the differentiation of equation (2.52) and using equations (2.2), (2.3), (2.39), (2.47) and (2.49) we can show that (and the reader should verify this for himself)

$$E = kT^2 \left(\frac{\partial Z_{\text{BE/FD}}}{\partial T}\right)_{\alpha,V} \qquad (2.54)$$

$$S = kZ_{\text{BE/FD}} + \frac{E}{T} \qquad (2.55)$$

$$p = kT \left(\frac{\partial Z_{\text{BE/FD}}}{\partial V}\right)_{\alpha,T} \qquad (2.56)$$

Statistical Thermodynamics of the Thermally Perfect Gas 57

Let us prove, for example, equation (2.56). From equation (2.44) we have

$$p = -\sum_n N_n^* \left(\frac{\partial \varepsilon_n}{\partial V}\right)_{N_n^*}$$

Consider

$$\left(\frac{\partial Z_{\text{BE/FD}}}{\partial V}\right)_{\alpha,T} = \sum_n g_n \frac{\partial}{\partial V} \ln\left[1 \pm \exp(\alpha)\exp\left(-\frac{\varepsilon_n}{kT}\right)\right]$$

$$= \sum_n g_n \left\{\frac{\exp(\alpha)\exp\left(-\frac{\varepsilon_n}{kT}\right)}{1 \pm \exp(\alpha)\exp\left(-\frac{\varepsilon_n}{kT}\right)} \cdot \frac{\partial}{\partial V}\left(\frac{\varepsilon_n}{kT}\right)\right\}$$

i.e.

$$\left(\frac{\partial Z_{\text{BE/FD}}}{\partial V}\right)_{\alpha,T} = \frac{1}{kT}\sum_n g_n \cdot \frac{1}{\exp\left(\frac{\varepsilon_n}{kT} - \alpha\right) \pm 1}\left(-\frac{\partial \varepsilon_n}{\partial V}\right)$$

$$= \frac{1}{kT}\sum_n g_n \cdot \frac{N_n^*}{g_n}\left(-\frac{\partial \varepsilon_n}{\partial V}\right)$$

$$= \frac{p}{kT}$$

which proves equation (2.56). We shall, in later chapters, use these partition functions to obtain the thermodynamic properties of various thermodynamic systems of interest to electrical engineers and physicists. We note, however, that before we can use the partition function quantitatively we must know the values of the g_n, ε_n and α.

2.13 The Maxwell–Boltzmann partition function and the evaluation of β

So far we have given two criteria for the validity of M.B. statistics, viz. equation (2.24)

$$N_n^* \ll g_n$$

and equation (2.40)

$$\exp(-\alpha)\exp\left(\frac{\varepsilon_n}{kT}\right) \gg 1$$

That these two criteria are the same is readily shown since equation (2.39) can be written

$$\frac{g_n}{N_n^*} \mp \begin{pmatrix}1\\ \text{or}\\ 0\end{pmatrix} = \exp(-\alpha)\exp\left(\frac{\varepsilon_n}{kT}\right)$$

where the minus sign is for B.E., the plus sign is for F.D. and the zero is for M.B. Clearly these formulae will be identical if

$$\frac{g_n}{N_n^*} \gg 1$$

which is equation (2.24). If the left-hand side is very much greater than one then so must be the right-hand side, i.e.

$$\exp(-\alpha) \cdot \exp\left(\frac{\varepsilon_n}{kT}\right) \gg 1$$

which is equation (2.40).

We have already noted in Section 2.6 that M.B. statistics applied when there were very many more translational quantum states available for occupation than there were particles to fill them. We can get some idea of how big this 'discrepancy' is by considering the helium atom (already discussed in Section 2.4) which moved freely throughout a container of volume one litre. We showed that the number of translational energy states accessible to the atom was about 10^{28} [see equation (2.13)]. If this container were filled with helium at STP then it would hold about 10^{22} particles. Hence the average number of particles in each quantum state is 10^{22} divided by 10^{28} or 10^{-6}, thus about one translational energy state in 1,000,000 has a particle in it and all the rest of the states are empty! This is why we called M.B. particles the 'few in number' particles in Sections 2.6 and 2.9.

Let us now find an expression for the M.B. partition function. This can be deduced from the generalized partition function for B.E. and F.D. statistics given by equation (2.52) viz.

$$Z_{\text{BE/FD}} = \pm \sum_n g_n \ln\left[1 \pm \exp(\alpha) \cdot \exp\left(-\frac{\varepsilon_n}{kT}\right)\right]$$

However, by taking the reciprocal of equation (2.40) we have for M.B. particles

$$\exp(\alpha) \exp\left(-\frac{\varepsilon_n}{kT}\right) \ll 1$$

whence, since $\ln(1+x) \sim x$ if x is small then we obtain

$$Z_{\text{BE/FD}} \sim \Sigma g_n \exp(\alpha) \exp\left(-\frac{\varepsilon_n}{kT}\right)$$

$$= \exp(\alpha) \Sigma g_n \exp\left(-\frac{\varepsilon_n}{kT}\right) \tag{2.57}$$

Statistical Thermodynamics of the Thermally Perfect Gas

Let us define,

$$Z_{\text{MB}} = \Sigma g_n \exp\left(-\frac{\varepsilon_n}{kT}\right) \tag{2.58}$$

Therefore

$$Z_{\text{BE/FD}} \rightarrow [\exp(\alpha)] Z_{\text{MB}}$$

if

$$\exp(-\alpha) \cdot \exp\left(\frac{\varepsilon_n}{kT}\right) \gg 1$$

Equations (2.53), (2.54), (2.55) and (2.56) may be shown to reduce to (and the reader should prove this for himself)

$$N = [\exp(\alpha)] Z_{\text{MB}} \tag{2.59}$$

$$E = NkT^2 \left(\frac{\partial}{\partial T} \ln Z_{\text{MB}}\right)_V \tag{2.60}$$

$$S = \frac{E}{T} + Nk \ln\left(\frac{e Z_{\text{MB}}}{N}\right) \tag{2.61}$$

where e is the exponential; also

$$p = NkT \left(\frac{\partial}{\partial V}\right) (\ln Z_{\text{MB}})_T \tag{2.62}$$

We have, in Section 2.10, assumed that $\beta = -\frac{1}{kT}$. The reader will recall that β was introduced in equation (2.36) as an undetermined multiplier with units of reciprocal energy. Let us now prove that $\beta = -\frac{1}{kT}$. We start with equations (2.38) and (2.40) for M.B. particles

$$\frac{N_n^*}{g_n} = \exp(\alpha) \exp(\beta \varepsilon_n)$$

Applying equation (2.2) gives

$$N = \sum_n N_n^* = \sum_n [g_n \exp(\alpha) \exp \beta \varepsilon_n]$$

$$= \exp(\alpha) \cdot [\sum_n g_n \exp \beta \varepsilon_n]$$

$$= [\exp \alpha] Z_{\text{MB}} \qquad \text{using equation (2.58)}$$

Therefore,

$$\exp \alpha = \frac{N}{Z_{\text{MB}}} \tag{2.63}$$

Therefore, equation (2.39) becomes

$$\frac{N_n^*}{g_n} = \frac{N \exp \beta \varepsilon_n}{Z_{\text{MB}}} \tag{2.64}$$

which is known as the Boltzmann distribution law.

We have shown in Section 2.4 that the translational energy levels of any thermally perfect gas are so closely spaced that we can treat them as being essentially continuous, thus we can effectively ignore quantization. We shall therefore write equation (2.64) in a form more appropriate for a *continuous* distribution of energy states. To do this we need to make four changes

a. Replace ε_n by the continuous variable ε
b. Replace N_n^* by $N(\varepsilon)d\varepsilon$ which is the number of particles having energy between ε and $\varepsilon + d\varepsilon$ for the most probable macrostate
c. Replace g_n by $g(\varepsilon)d\varepsilon$ which is the density of states function or the number of energy states lying between energies ε and $\varepsilon + d\varepsilon$ already derived in equation (2.14) viz.

$$g(\varepsilon)\, d\varepsilon = 2\pi (2m)^{\frac{3}{2}} V h^{-3} \varepsilon^{\frac{1}{2}}\, d\varepsilon \tag{2.14}$$

d. Replace the M.B. partition function by an integral rather than a summation.

$$Z_{\text{MB}} = \sum_n g_r \exp(\beta \varepsilon_n)$$

$$\rightarrow \int_0^\infty g(\varepsilon) \exp(\beta \varepsilon) d\varepsilon$$

i.e.

$$Z_{\text{MB}} = 2\pi(2m)^{\frac{3}{2}} V h^{-3} \int_0^\infty \varepsilon^{\frac{1}{2}} \exp(\beta \varepsilon) d\varepsilon$$

$$= 2\pi(2m)^{\frac{3}{2}} V h^{-3} \left(-\frac{1}{\beta}\right)^{\frac{3}{2}} \int_0^\infty x^{\frac{1}{2}} \exp(-x)\, . \, dx$$

where we have substituted $x = -\beta \varepsilon$

This last integral is a standard form and has the value $\dfrac{\sqrt{\pi}}{2}$

Statistical Thermodynamics of the Thermally Perfect Gas

Hence

$$Z_{\text{MB}} = \frac{\sqrt{\pi}}{2} 2\pi (2m)^{\frac{3}{2}} V h^{-3} \left(-\frac{1}{\beta}\right)^{\frac{3}{2}} \tag{2.65}$$

Equation (2.64) thus becomes

$$N(\varepsilon) d\varepsilon = 2\pi (2m)^{\frac{3}{2}} V h^{-3} \varepsilon^{\frac{1}{2}} d\varepsilon \cdot N \frac{2}{\sqrt{\pi}} \cdot \frac{1}{2\pi(2m)^{\frac{3}{2}} V h^{-3}} \cdot \frac{\exp \beta\varepsilon}{\left(-\frac{1}{\beta}\right)^{\frac{3}{2}}}$$

$$= \frac{2N \cdot \varepsilon^{\frac{1}{2}} \exp(\beta\varepsilon) d\varepsilon}{\left(-\frac{1}{\beta}\right)^{\frac{3}{2}} \sqrt{\pi}} \tag{2.66}$$

This is the Maxwellian distribution of translational energies of a thermally perfect gas. Since the reader no doubt is used to seeing it in terms of particle speeds rather than translational energies we can put $\varepsilon = \frac{1}{2}mv^2$ and obtain

$$N(v) dv = N \sqrt{\frac{2}{\pi}} (-\beta m)^{\frac{3}{2}} \cdot v^2 \exp(\beta m \cdot \frac{1}{2}v^2) dv \tag{2.67}$$

This should be compared with the usual result obtainable from kinetic theory

$$N(v) dv = N \sqrt{\frac{2}{\pi}} \left(\frac{m}{kT}\right)^{\frac{3}{2}} v^2 \exp\left(-\frac{mv^2}{2kT}\right) dv \tag{2.68}$$

whence

$$\beta = -\frac{1}{kT} \tag{2.69}$$

Although this analysis gives a value for β for a thermally perfect gas obeying M.B. statistics the result is true for all three statistics in the thermally perfect gas approximation. The reason is that β was introduced as a common undetermined multiplier for all three classes of statistics and once it is determined for one class it is determined for them all.

If we substitute this value of β back into equation (2.65) then we obtain

$$Z_{\text{MB}} = \frac{(2\pi mkT)^{\frac{3}{2}}}{h^3} \cdot V \tag{2.70}$$

which is a result that we will use several times in this book—it is the translational partition function of a thermally perfect gas of M.B. particles. We can use this result to frame yet another criterion for the applicability of M.B. statistics. We have already quoted two criteria at the beginning of this section and the one that we shall now derive is more readily amenable to direct calculation. Equation (2.40) is the criterion for M.B. statistics to apply, namely

$$\exp(-\alpha)\exp\left(\frac{\varepsilon_n}{kT}\right) \gg 1$$

Since $\frac{\varepsilon_n}{kT}$ is always positive or zero then the inequality is certainly satisfied if

$$\exp(-\alpha) \gg 1 \qquad (2.71)$$

But we have shown in equations (2.57) and (2.58), in the M.B. approximation

$$Z_{\text{BE/FD}} = \exp(\alpha)[Z_{\text{MB}}]$$

i.e.

$$Z_{\text{BE/FD}} = \exp(\alpha) \cdot \left[\frac{(2\pi mkT)^{\frac{3}{2}}}{h^3}V\right] \qquad (2.72)$$

But from equation (2.59)

$$N = \exp(\alpha)[Z_{\text{MB}}]$$
$$= Z_{\text{BE/FD}} \qquad (2.73)$$

where the latter result comes from equations (2.57) and (2.58). If we combine equations (2.72) and (2.73), then we get

$$N = \exp(\alpha)\left[\frac{(2\pi mkT)^{\frac{3}{2}}V}{h^3}\right]$$

i.e.

$$\exp(-\alpha) = \frac{(2\pi mkT)^{\frac{3}{2}}}{h^3}\frac{V}{N} \qquad (2.74)$$

If we combine equations (2.71) and (2.74) then we obtain the required criterion for M.B. statistics.

$$\frac{(2\pi mkT)^{\frac{3}{2}}}{h^3} \cdot \frac{V}{N} \gg 1 \qquad (2.75)$$

An examination of equation (2.75) shows that a thermodynamic system composed of N particles of a thermally perfect gas will obey M.B. statistics provided

a. T is large
b. m is large
c. $\frac{N}{V}$ i.e. the gas density is small

Statistical Thermodynamics of the Thermally Perfect Gas

The left-hand side of equation (2.75) is called the degeneracy parameter and is given the symbol Λ. Let us evaluate the degeneracy parameter for helium gas at STP filling a volume of one litre using numerical data from Appendix 1.

$$N = \frac{6 \cdot 02 \times 10^{26}}{22 \cdot 4} \times 10^{-3} = 2 \cdot 69 \times 10^{22} \text{ particles}$$

$$V = 10^{-3} m^3$$

$$m = 6 \cdot 64 \times 10^{-27} \text{ kg}$$

$$h = 6 \cdot 63 \times 10^{-34} \text{ J s}$$

$$k = 1 \cdot 38 \times 10^{-23} \text{ J }^\circ\text{K}^{-1}$$

$$T = 273 \,^\circ\text{K}$$

whence

$$\Lambda \sim 10^5$$

which clearly satisfies our inequality. In fact the inequality can be shown to be satisfied by *all* laboratory gases up to their highest densities (the Critical Point beyond which they are essentially more liquid than gas). Hence, all laboratory gases obey M.B. statistics. This is a very important conclusion because as we have seen, we can obtain all the thermodynamic properties from the appropriate partition function and we have already derived the translational partition function—equation (2.70). (In Chapter 3 we shall derive the partition function of an M.B. gas composed of particles possessing internal structure.) Since it is generally true that the determination of partition functions is much easier for M.B. statistics than for B.E. and F.D. statistics, then purely from the point of view of obtaining answers easily, M.B. statistics is preferable to either of the other two. As a consequence it is fortunate that all laboratory gases obey M.B. statistics!

However, when we come to consider the thermally perfect valence-electron gas in a metal which is discussed extensively in Chapter 4 then we find that our inequality is *not* satisfied. We can show this quite readily using the assumption that each metal atom gives one valence electron to the valence band, for instance in the case of copper.

$$\frac{V}{N} \sim 1 \cdot 18 \times 10^{-29} \text{ m}^3 \text{ per free electron}$$

whence at $T = 273\,^\circ\text{K}$ our degeneracy parameter has the value $1 \cdot 3 \times 10^{-4}$, that is, it is very much less than unity and our inequality is not satisfied. *We therefore cannot treat the valence-electron gas by M.B. statistics.* The reason is twofold:

a. The low electron mass compared with the mass of laboratory gas particles

b. The high electron density in a metal such as 10^{28} electrons per cubic metre compared with a value of about 10^{25} particles per cubic metre for laboratory gases at STP.

The reader who knows something about ionized gases may wonder whether a laboratory plasma behaves as an F.D. or an M.B. gas. The answer is that the free-electron densities achievable in laboratory plasmas seldom approach the values that they have in metals; rather, they usually range between 10^{17} to 10^{24} electrons per cubic metre. Consequently laboratory-ionized plasmas behave as thermally perfect M.B. gases and we shall analyse their behaviour on this basis in the next chapter.

2.14 The macroscopic and microscopic consequences of placing two thermodynamic systems in contact

Chapter 1 has been concerned with a brief review of the thermodynamic properties of a system (usually closed) and we have shown in Section 2.11 how the macroscopic equations of Chapter 1 according to the First and Second Laws can be related to the microscopic description of the same system developed in the present chapter. Before leaving the macroscopic equations and concerning ourselves solely with the physical aspects of the various types of thermally perfect gas considered in this book, let us consider the effects of placing any two thermodynamic systems of thermally perfect gases into contact. This is not just an academic exercise because there are many situations in electrical technology where such 'systems in contact' are very important. For example, the p–n junction between a p-type and an n-type semiconductor which is the essence of a transistor; we shall consider the p–n junction in Chapter 7. Or the contact between a metal and a semiconductor such as either occurs in the contact rectifier or in a thermoelectric device; or the contact between the electron gas in a metal and the electron gas in the evacuated region outside the metal such as occurs in the Richardson–Dushman thermionic emission process.

Generally speaking, when we place two thermodynamic systems in contact we have to define what is meant by 'contact'. In the case considered here we shall be quite general, that is, we shall allow the two systems to voluntarily exchange either energy or particles and to deform each other's system boundary if they want to. In Section 1.4 we have discussed the concept of the thermodynamic equilibrium between a system and its surroundings (which of course is just the same as between one system and another since the surroundings always constitute 'another system'). We showed that if the two systems were to be in thermodynamic equilibrium then this required:

1. Thermal equilibrium to exist between the two systems so that the temperature was the same throughout both systems.

Statistical Thermodynamics of the Thermally Perfect Gas 65

2. Mechanical equilibrium to exist, all forces exerted between the first system on the second system being uniform throughout the first system and exactly balanced by the external forces exerted by the second system on the first system.
3. Chemical equilibrium to exist, obviating any tendency for a net flux of particles from one system to another.

Let us see what information our statistical equations give us about the thermodynamic equilibrium of two systems in contact. We must be careful to note that either system is 'open' and not 'closed' because each system can accept not only heat and/or work transfers from the other system but also mass transfers (exchange of particles). Accordingly the thermodynamic equations derived in Chapter 1 for closed systems are too restrictive for the present discussion.

Equation (2.47) showed that for all three statistics for a closed system of thermally perfect gas

$$d(\ln \Omega) + \alpha \sum_n dN_n^* + \beta \sum_n \varepsilon_n dN_n^* = 0$$

Equation (2.49) gave

$$dS = k\,d(\ln \Omega)$$

Hence

$$dS + k\alpha \sum_n dN_n^* + k\beta \sum_n \varepsilon_n dN_n^* = 0 \qquad (2.76)$$

But from equation (2.43)

$$\Sigma \varepsilon_n dN_n^* = dE - \Sigma N_n^* d\varepsilon_n$$

whence equation (2.76) becomes

$$dS + k\alpha \Sigma dN_n^* + \beta k\,dE - \beta k \Sigma N_n^* d\varepsilon_n = 0 \qquad (2.77)$$

Now suppose that our system is such that the only work transfers that occur to and from it are displacement work transfers due to changes in the system volume—see Section 1.4; then from equations (2.44) and (2.45) the work-transfer term (remembering that the work was done *on* the system and is therefore negative) is

$$\mathrm{d}W = -p\,dV = \Sigma N_n^* d\varepsilon_n$$

Hence equation (2.77) becomes

$$dS + \alpha k \Sigma dN_n^* + \beta k\,dE + \beta k p\,dV = 0 \qquad (2.78)$$

Equation (2.69) gives

$$\beta = -\frac{1}{kT}$$

whence equation (2.78) becomes

$$dS + \alpha k \Sigma dN_n^* - \frac{dE}{T} - \frac{pdV}{T} = 0$$

which on rearrangement is

$$TdS - pdV = dE - \alpha kT \Sigma dN_n^* \qquad (2.79)$$

If we compare equations (2.79) and (1.11) we see that the former has an additional term, the second term on the right-hand side. If the system is closed so that no mass transfers are allowed between it and its surroundings then equations (2.2) and (2.33) hold, viz.

$$\Sigma dN_n^* = 0$$

so that equations (1.11) and (2.79) are now identical.

Let us now consider two systems A and B brought together such that they can exchange energy and particles and deform each other's system boundaries. Then the two systems considered together may be regarded as an isolated system AB. Now when A and B are placed together we did not require that both have the same temperature so that they can be initially in thermal non-equilibrium; nor did we require that they be in mechanical equilibrium because they can mutually change their system volumes as they expand or contract upon exchanging energy. Finally, they were not required to be in chemical equilibrium because they were free to exchange particles. However, after they had been placed in contact they eventually will reach thermodynamic equilibrium and it is this state of thermodynamic equilibrium that we shall now analyse. Thus we are not concerned with the initial thermodynamic states of the two systems A and B before they were put together; neither are we concerned with their intermediate states as they proceed towards thermodynamic equilibrium; all we are concerned with is their final state of thermodynamic equilibrium.

The First Law applied to the isolated combined system is, using equation (1.1)

$$đQ_{AB} - đW_{AB} = dE_{AB}$$

But if the system AB is isolated it can undergo no heat and/or work transfer, i.e.

$$đQ_{AB} = 0; \quad đW_{AB} = 0$$

i.e.

$$dE_{AB} = 0 \qquad (2.80)$$

Statistical Thermodynamics of the Thermally Perfect Gas

But
$$E_{AB} = E_A + E_B$$
∴
$$dE_{AB} = 0 = dE_A + dE_B$$
i.e.
$$dE_B = -dE_A \tag{2.81}$$

The Second Law applied to the isolated combined system, is, using equation (1.11)
$$T dS_{AB} - (p dV)_{AB} = dE_{AB}$$
But
$$(p dV)_{AB} = 0$$
and
$$dE_{AB} = 0$$
from equation (2.80)
∴
$$dS_{AB} = 0 \tag{2.82}$$
But
$$S_{AB} = S_A + S_B$$
∴
$$dS_{AB} = 0 = dS_A + dS_B$$
i.e.
$$dS_B = -dS_A \tag{2.83}$$

Also, since there are no unbalanced forces to expand or contract the combined system volume V_{AB} we have
$$dV_B = -dV_A \tag{2.84}$$
This, in fact, is the justification for writing $(p dV)_{AB} = 0$.

Now if we use equation (2.79) in equation (2.82) we obtain
$$dS_{AB} = \frac{dE_A}{T_A} + \frac{dE_B}{T_B} + \frac{p_A dV_A}{T_A} + \frac{p_B dV_B}{T_B}$$
$$- \alpha_A k (\sum_n dN_n^*)_A - \alpha_B k (\sum_n dN_n^*)_B = 0$$

Using equations (2.81), (2.82), (2.84), this reduces to
$$0 = \left(\frac{1}{T_A} - \frac{1}{T_B}\right) dE_A + \left(\frac{p_A}{T_A} - \frac{p_B}{T_B}\right) dV_A$$
$$- \alpha_A k (\sum_n dN_n^*)_A - \alpha_B k (\sum_n dN_n^*)_B \tag{2.85}$$

Now since the combined system AB is isolated it can neither receive or lose particles, i.e.

$$(\sum_n N_n^*)_A + (\sum_n N_n^*)_B = \text{Constant}$$

On differentiating this the equation becomes

$$(\sum_n dN_n^*)_A + (\sum_n dN_n^*)_B = 0 \tag{2.86}$$

If we substitute equation (2.86) into (2.85) we get

$$0 = \left(\frac{1}{T_A} - \frac{1}{T_B}\right) dE_A + \left(\frac{p_A}{T_A} - \frac{p_B}{T_B}\right) dV_A - (\alpha_A - \alpha_B)k(\sum_n dN_n^*)_A \tag{2.87}$$

In the state of thermodynamic equilibrium the only possible solution to equation (2.87) is

1. Thermal equilibrium exists, i.e. $T_A = T_B$
2. Mechanical equilibrium exists, i.e. $p_A = p_B$
3. Chemical equilibrium exists, i.e. $\alpha_A = \alpha_B$

It is this last result that is important in the present discussion, namely

$$\alpha_A = \alpha_B \tag{2.88}$$

where α is known as the chemical potential and is most used in chemical thermodynamics such as combustion theory, electrolyte theory and so on. We shall, however, use equation (2.88) in our discussion of metal and semiconductor junctions.

CHAPTER THREE

The Thermally Perfect Maxwell–Boltzmann Gas

3.1 The thermodynamic properties of an M.B. gas composed of monatomic atoms

In Chapter 2 we have shown how all of the thermodynamic properties of a thermally perfect gas obeying any of the three classes of statistics can be determined if we know the appropriate partition function. In this chapter we shall briefly discuss the thermodynamic properties of the thermally perfect M.B. gas comprised firstly of monatomic atoms such as He, Ne, Ar, etc., and secondly a mixture of monatomic atoms, ions and electrons, a partially ionized gas known as a plasma.

Of course we could apply M.B. statistics to diatomic or polyatomic gases but these applications are more of interest to the physical chemist or mechanical engineer. We shall, in later chapters, also use M.B. statistics to describe, for example, that special group of electrons called the conduction electrons in a metal or semiconductor. However, in these later applications we utilize M.B. statistics as a very special case of electron behaviour whereas in this chapter M.B. statistics are the norm for all 'ordinary' laboratory gases.

Consider the monatomic atom. We have already derived the translational partition function for an M.B. gas of structureless particles in equation (2.70)

$$Z^{\text{trans}} = \frac{(2\pi mkT)^{\frac{3}{2}} V}{h^3} \qquad (3.1)$$

In this section we shall consider the term 'structureless' when applied to a particle, to indicate that the particle can possess translational that is, kinetic energy only (in Section 3.2 we shall give an example of a particle with structure).

From equations (2.59) to (2.62) we obtain, using equation (3.1),

$$N = \exp\alpha \frac{(2\pi mkT)^{\frac{3}{2}}V}{h^3}$$

i.e.

$$\exp\alpha = (2\pi mkT)^{\frac{3}{2}} h^{-3} \frac{V}{N} \tag{3.2}$$

also

$$E = NkT^2 \left(\frac{\partial}{\partial T} \ln Z^{\text{trans}}\right)$$

$$= \tfrac{3}{2} NkT \text{ on differentiation} \tag{3.3a}$$

We can immediately use this result to obtain the average thermal kinetic energy per particle (the energy due to random motion)

$$\bar{\varepsilon} = \frac{E}{N} = \frac{3}{2} kT \tag{3.3b}$$

We have equation (2.61) for the entropy S,

$$S = \frac{E}{T} + Nk \ln\left(\frac{eZ^{\text{trans}}}{N}\right) \quad \text{where } e \text{ is the exponential}$$

i.e.

$$S = \frac{3}{2} Nk + Nk \ln\left[\frac{e}{N} \frac{(2\pi mkT)^{\frac{3}{2}}}{h^3} \cdot V\right] \tag{3.4}$$

which is known as the Sackur–Tetrode equation.

We also have, for the pressure

$$p = NkT \frac{\partial}{\partial V}(\ln Z^{\text{trans}}) = \frac{NkT}{V} \tag{3.5}$$

The reader will immediately recognise this as the well-known equation of state of the 'perfect' or 'ideal' gas so much used by thermodynamicists!

Also, from equation (3.3) we can deduce that

Heat capacity at constant volume $C_V = \left(\frac{\partial E}{\partial T}\right)_V$

$$= \frac{3}{2} Nk \tag{3.6}$$

If we consider one mole of monatomic gas (see Appendix 2 for definition of a mole) then equation (3.6) becomes

$$\text{Molar heat capacity at constant volume } \bar{C}_V = \tfrac{3}{2}R \qquad (3.7)$$

since $kN_{\text{AV}} = R$ where R is the Universal Gas Constant and N_{AV} the Avogadro number.

This is the value obtained in laboratory experiments. It is also the value predicted by simple kinetic theory developed long before energy quantization had been discovered. Of course, we have obtained our result by effectively ignoring quantization, so the agreement is not surprising!

The reader may wonder, in fact, where quantization is important. The answer to this is quite straightforward and has been given in Sections 2.4 and 2.13. The *translational* energy levels of all particles (bosons or fermions) of thermally perfect gases in 'laboratory-sized' containers although quantized, are in practise consistently so closely spaced that their *discrete separation can always be ignored*. Of course, it is for precisely this reason that energy quantization remained undiscovered for several centuries! Energy quantization remains apparent and must *not* be 'smeared out' *only* if the particles are restricted to atomic-sized volumes—or if they possess internal structure (which is another way of saying the same thing). As an example of a particle possessing internal structure a diatomic molecule can be thought of as two atoms connected together by a spring. The entire molecule can move with translational energy throughout its container and, as we have already remarked, we can quite justifiably treat its quantized translational energy levels as having an essentially continuous distribution if this container is of laboratory size. However, the molecule can also rotate 'end over end' and therefore possesses rotational energy; it can also vibrate along the line of centres of the two atoms so that the molecule can also possess vibrational energy. Both rotational and vibrational energy are quantized; an analysis shows that the quantization of rotational energy is 'fine', i.e. the rotational energy levels are quite closely spaced— (but nothing as closely spaced as the translational energy levels); their discrete nature may be observed spectroscopically and accurate thermodynamic calculations of the rotational heat capacities of molecules requires us to take these quantized rotational states into account. The vibrational energy levels are quite widely spaced for most common diatomic molecules and vibrational quantization is readily observable spectroscopically. In this case we cannot treat the vibrational energy levels as continuously distributed. Vibrational quantization has a considerable importance thermodynamically, especially in heat-capacity problems for diatomic gases at both high and low temperatures.

As we have remarked, quantization is important when the particle is confined to an atomic sized volume; let us consider an important application of this.

3.2 The partition function of a particle possessing electronic structure

Consider any electron bound to its parent nucleus in an atom. Here the electron is free 'to move in its Bohr orbit around the nucleus' but cannot stray too far from the nucleus since it is bound to it. As a consequence its quantized energy levels show a very wide separation and cannot be considered as continuously distributed. This readily follows from equation (2.9), where we have shown that for a particle in a container of volume V the quantized translational energy values are inversely proportional to the two-thirds power of the volume. Hence from equation (2.15) if the volume of the 'container' is infinitesimal (as it is for the electron bound to its nucleus) then the average energy separation between successive energy levels will be large. These discrete electronic levels are readily detectable spectroscopically.

Consider the monatomic atom that we have just discussed. We can think of such an atom as possessing:

a. Quantized energy of translation as it moves around its container. This is the energy which we have shown to be essentially continuously distributed if the container is of laboratory size.
b. The quantized energy of electronic excitation, if any electron of the atom is not in its ground electronic level (the energy level of lowest energy).

We can assume that these two types of energy are not coupled in any way, and that the atom is free to move randomly throughout its laboratory-sized container regardless of its state of electronic excitation and vice versa. We can therefore write the total energy of the atom as the sum of these two energies,

$$\varepsilon_{\text{total}} = \varepsilon_n^{\text{trans}} + \varepsilon_e^{\text{int}} \tag{3.8}$$

Here

$\varepsilon_n^{\text{trans}}$ = translational energy of the atom characterized by a translational quantum number n

and

$\varepsilon_e^{\text{int}}$ = the internal energy of the atom, thus the electronic excitation energy in an excited electronic level characterized by one (or perhaps a set) of internal quantum numbers e.

We shall deliberately not use the correct, accepted symbols for internal quantum numbers since these might be confused with the symbols already introduced as quantum numbers.

The M.B. partition function has been given by equation (2.58), for the case of translational energy. In fact the expression is quite general and we can write,

$$Z_{\text{MB}} = \Sigma g_m \exp\left(-\frac{\varepsilon_m}{kT}\right)$$

where m refers to the quantum number or numbers appropriate to the problem and ε_m refers to the sum of the various types of energy. For the particular case we are considering of a monatomic atom with excited electronic states this is

$$Z_{\text{MB}} = \sum_n \sum_e g_n g_e \exp\left(-\frac{\varepsilon_n^{\text{trans}} + \varepsilon_e^{\text{int}}}{kT}\right) \tag{3.9}$$

$$= \left[\sum_n g_n \exp\left(-\frac{\varepsilon_n^{\text{trans}}}{kT}\right)\right]\left[\sum_e g_e \exp\left(-\frac{\varepsilon_e^{\text{int}}}{kT}\right)\right]$$

i.e.
$$Z_{\text{MB}} = Z^{\text{trans}} \cdot Z^{\text{int}} \tag{3.10}$$

Equation (3.10) shows that the total M.B. partition function is the product of the separate partition functions that describe the two independent modes of

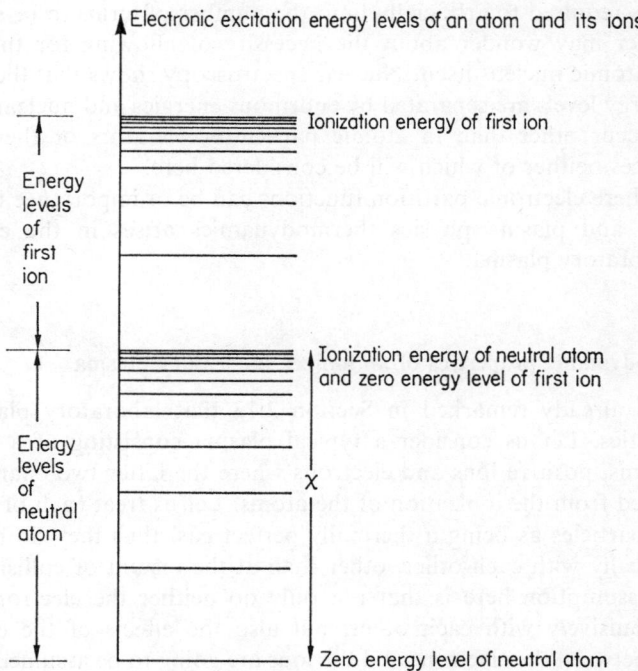

Figure 3.1. Schematic representation of the 'ladder' of allowed electronic excitation energy levels of a neutral atom and its first ion. Note that the most tightly bound level (of electron to nucleus) is taken as the zero energy level of the neutral atom; this is a spectroscopic convention.

energy, translational and electronic excitation energy. We have already shown how to evaluate the translational part of equation (3.10). Let us consider

$$Z^{\text{int}} = \sum_e g_e \exp\left(-\frac{\varepsilon_e^{\text{int}}}{kT}\right) \tag{3.11}$$

To evaluate this sum we need to know the values of the energies of the various electronic levels of electrons in monatomic atoms. These are determinable spectroscopically and also calculable using quantum mechanics but only for simple monatomic atoms; they are shown schematically in Figure 3.1. We shall not, however, discuss these quantized electronic levels because it would lead us too far into the vector model of the atom which is far removed from thermodynamics! It is sufficient to note that the right-hand side of equation (3.11) cannot be summed algebraically and is just usually left as a series. This series is 'well-behaved' at medium temperatures, thus convergent but unfortunately tending to diverge at very high temperatures, so that some energy-state cut-off procedure is required to artificially limit the number of terms to be summed.

The reader may wonder about the necessity of allowing for the structure within the atomic nucleus itself. Nuclear spectroscopy shows that the quantized nuclear energy levels are separated by enormous energies and nuclear excitation does not occur other than in atomic particle accelerators or thermonuclear fusion devices neither of which will be considered here.

A case where electronic partition functions can be of importance in electrical engineering and plasma physics thermodynamics arises in the case of the ionized-laboratory plasma.

3.3 Thermodynamic properties of an ionized laboratory plasma

We have already remarked in Section 2.13 that laboratory plasmas obey M.B. statistics. Let us consider a typical plasma consisting of a mixture of neutral atoms, positive ions and electrons where the latter two charged species have resulted from the ionization of the atoms. Let us treat each of these three species of particles as being a thermally perfect gas, thus they do not interact electrostatically with each other, other than at the instant of collision. Clearly the main assumption here is that not only do neither the electrons nor ions interact repulsively with each other, but also the effects of the electrostatic attraction between the electrons and the ions are going to be assumed negligible. Rather surprisingly this is a very fair approximation for most laboratory plasmas and only fails when the free-electron densities become about 10^{24} particles per cubic metre or greater, which is the situation existing in a metal as we shall show in Chapter 4.

The Thermally Perfect Maxwell–Boltzmann Gas

We can therefore consider our plasma as three non-interacting gases occupying the same container. On this basis we can write for the partition functions

$$Z_{\text{atom}} = (2\pi m_{\text{atom}} kT)^{\frac{3}{2}} h^{-3} V Z^{\text{int}}_{\text{atom}} \tag{3.12}$$

$$Z_{\text{ion}} = (2\pi m_{\text{ion}} kT)^{\frac{3}{2}} h^{-3} V Z^{\text{int}}_{\text{ion}} \tag{3.13}$$

$$Z_{\text{electron}} = (2\pi m_e kT)^{\frac{3}{2}} h^{-3} V . 2 \tag{3.14}$$

where m_e = mass of the electron.

In equation (3.14) the 2 on the right-hand side represents the two allowed spin quantum states of the free electron, a point which we shall reconsider in Chapter 4. In the first two equations the Z^{int} represents the internal partition function of the atom and ion and, as we have seen in the previous section, these allow for the fact that both species may possess internal structure in the form of occupied excited electronic levels.

Let there be N_a atoms, N_i ions and N_e electrons present in the container. Since a plasma is essentially electrically neutral we must have

$$N_i = N_e \tag{3.15}$$

assuming that each ion has only one missing electron. Of course for all atoms other than hydrogen the possibility of multiple ionization exists with the loss of more than one electron. However, this only takes place at very high plasma temperatures, say above 10,000°K and low plasma pressures.

Since our three species are non-interacting, their total energies are completely separable and we can write

$$E = E_a + E_i + E_e = N_a kT^2 \frac{\partial}{\partial T}(\ln Z_a)_V + N_i kT^2 \frac{\partial}{\partial T}(\ln Z_i)_V$$

$$+ N_e kT^2 \frac{\partial}{\partial T}(\ln Z_e)_V \tag{3.16}$$

using equation (2.60). The equation of state of the plasma is given by equation (2.62) and is

$$p = N_a kT \frac{\partial}{\partial V}(\ln Z_a)_T + N_i kT \frac{\partial}{\partial V}(\ln Z_i)_T + N_e kT \frac{\partial}{\partial V}(\ln Z_e)_T \tag{3.16a}$$

Let us combine equations (3.12) through (3.14) with equation (3.16) remembering that

$$NkT^2 \frac{\partial}{\partial T}(\ln Z) = \frac{3}{2} NkT + NkT^2 \frac{\partial}{\partial T}(\ln Z^{\text{int}}) \tag{3.17}$$

and obtain the result for the thermodynamic energy of the plasma

$$E = \frac{3}{2}kT[N_a + N_i + N_e] + kT^2\left[N_a\frac{\partial}{\partial T}(\ln Z_a^{\text{int}})_V\right]$$
$$+ kT^2\left[N_i\frac{\partial}{\partial T}(\ln Z_i^{\text{int}})\right] + N_i\chi \qquad (3.18)$$

Here the last term on the right-hand side arises from the fact that the electronic energy levels of the ion start at the ionization potential χ of the neutral atom. This is shown in Figure 3.1. Accordingly we must incorporate into our analysis the fact that *each* electron–ion pair is 'carrying' the ionization energy χ and since there are N_i ion-pairs, these 'carry' a total energy $N_i\chi$.

Also, remembering that

$$NkT\frac{\partial}{\partial V}(\ln Z)_T = NkT\frac{\partial}{\partial V}\left[\ln\frac{(2\pi mkT)^{\frac{3}{2}} \cdot V}{h^3} \cdot Z^{\text{int}}\right] = NkT\left(\frac{1}{V}\right)$$

since the internal partition function of an atom or ion depends only upon the temperature of the gas and the electronic levels of the particle and not upon the laboratory volume within which the particle is contained. We obtain the equation of state

$$p = \frac{kT}{V}[N_a + N_i + N_e] \qquad (3.19)$$

It is usual in plasma theory to introduce the degree of ionization α where

$$\alpha = \frac{\text{number of electrons or ions present}}{\text{number of atoms originally present before ionization occurred}}$$
$$= \frac{N_i}{N_a + N_i} = \frac{N_i}{N} \qquad (3.20)$$

Here N = total number of atoms originally present before ionization
Then,
$$\text{number of atoms} = N_a = (1 - \alpha)N$$
$$\text{number of electrons} = N_e = \text{number of ions}$$
i.e
$$N_e = \alpha N$$
whence equations (3.18) and (3.19) become

$$E = NkT\left[\frac{3}{2}(1 + \alpha)\right] + N\alpha\chi + NkT^2(1 - \alpha)\frac{\partial}{\partial T}(\ln Z_a^{\text{int}})$$
$$+ NkT^2\alpha\frac{\partial}{\partial T}(\ln Z_i^{\text{int}})$$

The Thermally Perfect Maxwell–Boltzmann Gas

and

$$pV = (1 + \alpha)NkT \tag{3.24}$$

Other thermodynamic properties can be deduced; for example the heat capacity at constant volume is

$$C_V = \left(\frac{\partial E}{\partial T}\right)_V = \frac{3}{2} Nk(1 + \alpha) + NkT\left(\frac{3}{2} + \frac{\chi}{kT}\right)\left(\frac{\partial \alpha}{\partial T}\right)_V$$

+ derivatives with respect to temperature of the terms involving the internal partition functions. (3.25)

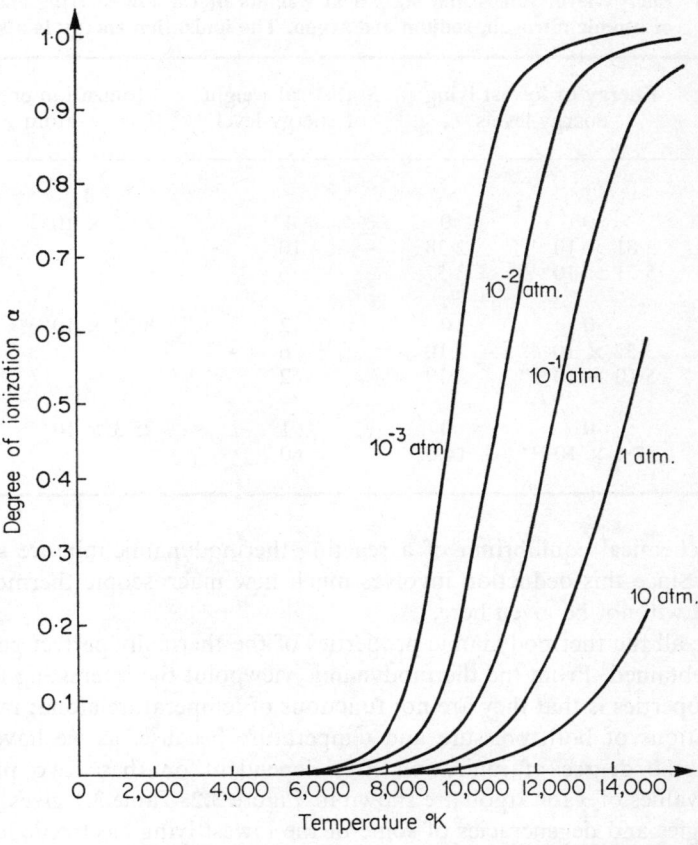

Figure 3.2 Degree of ionization of argon α versus temperature plotted for various plasma total pressures.

We should note that so far, although we have defined the degree of ionization, we have given no method of finding it—or its derivatives. In fact, α is determined by *both* the pressure p and the temperature T of the plasma and is given by,

$$\left(\frac{\alpha^2}{1-\alpha^2}\right) \cdot \frac{p}{kT} = \frac{(2\pi m_e kT)^{\frac{3}{2}}}{h^3} \frac{2Z_i^{int}}{Z_a^{int}} \exp\left(-\frac{\chi}{kT}\right) \tag{3.26}$$

This quadratic equation for α is known as the Saha equation and can be deduced entirely from statistical thermodynamics. However, its deduction involves a thermodynamic property called the Gibbs function, together with the criterion

Table 3.1 Energy-level values and statistical weights of the lowest lying energy levels of atomic nitrogen, sodium and argon. The ionization energy is also given.

Element	Energy of lowest lying energy levels		Statistical weight of energy level	Ionization energy of atom χ	
	J	eV		J	eV
N	0	0	4	$23 \cdot 3 \times 10^{-19}$	14·6
	$3 \cdot 81 \times 10^{-19}$	2·38	10		
	$5 \cdot 71 \times 10^{-19}$	3·57	6		
Na	0	0	2	$8 \cdot 22 \times 10^{-19}$	5·14
	$3 \cdot 37 \times 10^{-19}$	2·10	6		
	$5 \cdot 10 \times 10^{-19}$	3·19	2		
A	0	0	1	$25 \cdot 3 \times 10^{-19}$	15·8
	$22 \cdot 4 \times 10^{-19}$	14·0	60		

for the chemical equilibrium of a reacting thermodynamic mixture such as a plasma. Since this deduction involves much new macroscopic thermodynamic theory it will not be given here.

Hence all the thermodynamic properties of the thermally perfect gas plasma can be obtained. From the thermodynamic viewpoint the interesting feature of these properties is that they are not functions of temperature alone; rather they are functions of both pressure and temperature because, as we have just remarked, the degree of ionization α is dependent on these two properties. Typical values of α for argon are shown in Figure 3.2. Table 3.1 gives values of the energies and degeneracies of some of the lowest lying electronic levels of a few atoms; ionization energies are also given. Such energy-level data is necessary to determine the electronic partition function introduced in equation (3.11).

The Thermally Perfect Maxwell–Boltzmann Gas

Problems for Chapter 3

1. A laboratory container is filled with N atoms of a thermally perfect Maxwell–Boltzmann gas where each atom possesses both translational and electronic excitation energy. If the internal, that is electronic partition function of each atom can be approximated by the expression

$$g_0 + g_1 \exp - \frac{\varepsilon_1}{kT}$$

show that the constant-volume heat capacity of the gas is

$$\frac{3}{2} kN + kN \left(\frac{\varepsilon_1}{kT}\right)^2 \cdot \frac{\frac{g_1}{g_0} \exp\left(-\frac{\varepsilon_1}{kT}\right)}{\left\{1 + \left(\frac{g_1}{g_0}\right) \exp\left(-\frac{\varepsilon_1}{kT}\right)\right\}^2}$$

Show that at both high and low temperatures the constant-volume heat capacity of the gas is essentially that due to translational energy alone and explain in terms of the electronic energy-level occupation numbers why the electronic excitation contribution to the heat capacity for these temperature limits is essentially zero.

2. A closed rigid container is partitioned into two unequal volumes V_1 and V_2. Volume V_1 contains a thermally perfect Maxwell–Boltzmann gas made up of N atoms possessing no internal structure, at a pressure p_1 and temperature T; volume V_2 contains the same thermally perfect Maxwell–Boltzmann gas also of N atoms at the same temperature T but at a different pressure p_2.

If the partition is removed and the gases diffuse into each other, show that the entropy change in the diffusion process is

$$kN \ln \frac{(p_1 + p_2)^2}{4 p_1 p_2}$$

If the initial pressures as well as the initial temperatures had been the same, then the above formula shows that the change of entropy is zero. Explain how a spontaneous process like diffusion can occur with no entropy change (Gibbs' paradox).

CHAPTER FOUR

The Thermally Perfect Fermi-Dirac Electron Gas in a Metal

4.1 Electron energy bands in a metal (Sommerfeld model)

In this Chapter we shall examine the thermodynamic aspects of a simple theoretical model for electrons in a metal. In particular we shall restrict our thermodynamic system to contain just those electrons which are involved in the conduction of electrical charge (current flow) and thermal energy (heat flow). We shall see that these electrons represent a small fraction of all of the electrons in the metal; but they are a very important fraction! Any solid is comprised of one or more crystals where, in each crystal, the atoms are arranged in a more or less regular lattice. The actual lattice structure depends largely on the way the individual atoms of the crystal are bound together. There are in fact several types of crystal binding such as covalent binding, ionic binding, etc.; however, we shall not consider these types of crystal since they contain electrons which are so tightly bound that an externally applied electric field does not readily produce a flow of current. We shall concern ourselves solely with that form of crystal binding producing metals and in particular we shall be most interested in the alkali metals (Li, Na, K, Cs, Rb) and the noble metals (Cu, Ag, Au, etc.).

The simplest picture of a metal is, as we have remarked above, a lattice structure of atoms held together by binding forces. The innermost electrons of the lattice atoms always remain bound to their parent atom and 'move' around their parent nuclei in the same 'Bohr orbits' much as they would if each atom were completely isolated and not bound to its neighbours in the lattice. However, the outermost electrons, especially those in the incomplete outermost electron shell (the valence electrons) behave very differently. The reason is that their Bohr orbits are about the same size as the distance between the nuclei, so that the electrons of neighbouring atoms interact powerfully with each other. In fact each valence electron moves in the electrostatic field of its positive parent nucleus together with the electrostatic field produced by

The Thermally Perfect Fermi–Dirac Electron Gas

the inner-bound electrons of the same nucleus together with the 'averaged-out' electrostatic field of all those other valence electrons of other atoms interacting with it. What is the result of this interaction of each valence electron with these three electrostatic fields?

The result is that the individual electronic quantum levels of the valence electrons split into an energy band known as the valence band. This band consists of a very large number of very closely spaced quantum levels. The number of energy levels in this valence band equals the number of atoms in the metal and is therefore very large indeed; if our crystal is made up of N atoms then every valence-electron energy state of each atom combines with the identical state of $(N-1)$ identical atoms resulting in the formation of a valence band of N energy levels, all of which are very closely spaced.

This means that the valence electrons lead a communal existence in the valence band; they are no longer attached to any particular nucleus and are free to move throughout the entire volume of the metal. Since these valence electrons are constrained to move in a 'container', i.e. the metal, then their energies must be quantized. We can therefore see that there is a very close analogy between the translational energy states of a particle moving in a laboratory container (such as we discussed in Sections 2.3 and 2.4) and the translational energy states of a valence electron moving in the valence band of a metal since the electron can only possess translational, i.e. kinetic energy.

The reader may wonder why banding occurs. The answer has already been given in Section 2.6. Electrons are fermions and fermions must obey the Pauli principle. If banding within the metal did not occur then all of the valence electrons would want to move in the same valence energy state, and this is disallowed. The Pauli principle states that no two electrons can possess the same set of quantum numbers. But what are the quantum numbers that describe the valence electrons in a metal? In fact, each quantum state of a valence electron when in the valence band is characterized by just two quantum numbers, n and m_s. The quantum number n is the translational quantum number already introduced in Section 2.3; it must have an integral value and can reach very large numerical values—see equation (2.12). The quantum number m_s is the spin quantum number and can have only two values, $+\frac{1}{2}$ and $-\frac{1}{2}$.

We see then that two valence electrons *can* occupy the same translational energy state characterized by the translational quantum number n and not disobey the Pauli principle—*provided* that they have different values of spin quantum number; this means that one of the electrons must have $m_s = +\frac{1}{2}$ whilst the other electron must have $m_s = -\frac{1}{2}$. It is for this reason that in the first paragraph of Section 2.8 we referred to the 'single-particle translational energy state'. Such an energy state has prescribed n and m_s values and can only be occupied by one electron at most. We must take care to distinguish such a 'single-particle translational energy state' of an electron in the valence band of a metal from the 'translational energy state' used extensively in Chapter 2 because this latter state is prescribed by a given value of n only and therefore if we are considering the valence band of a metal, could contain up to two electrons provided these two electrons had different spin quantum numbers, $m_s = +\frac{1}{2}$ and $m_s = -\frac{1}{2}$. We shall, in the text, refer to both single-particle translational energy states' and also 'translational energy states', depending on the topic under discussion, and the reader should take care to distinguish between these two.

4.2 The thermally perfect degenerate electron gas in a metal

We have now described the electrons in a metal as being either bound to the parent nuclei if they are non-valence electrons (and we shall not consider such electrons any further in this chapter) or being free to move within the valence band if they were originally the valence electrons of the individual atoms. Although we have remarked above that the valence band is analogous to the closely spaced translational energy levels of a particle moving in a laboratory container, there are two differences to be considered. Firstly, we characterize the translational energy levels of the valence electrons in the valence band of a metal slightly differently from the discussion in Section 2.3. In that section we used the boundary conditions $\psi = 0$ at the walls of the container in order to obtain the allowed energy levels and wave functions of a particle in the container. This gave us periodic solutions for the wave function [see equation (2.10)], and if we made a closer examination of these solutions we would find that they represented a system of standing-matter waves in the container.

However, when considering the problem of free-valence electrons in a metal we do not use these fixed-end boundary conditions to obtain a set of standing waves. Rather, it proves more useful to use the method of periodic boundary conditions. Thus, quantize the electrons in the cube of side L and require that the wave functions be *periodic* in x, y, z with period L.

$$\psi(x + L, y, z) = \psi(x, y, z)$$

This gives *travelling* electron waves whose wave functions can readily be shown to be of the form

$$\psi_n = A \exp \frac{2\pi i}{L}(n_x . x + n_y . y + n_z . z)$$

where

$$n_x = 0, \pm 1, \pm 2, \ldots$$
$$n_y = 0, \pm 1, \pm 2, \ldots$$
$$n_z = 0, \pm 1, \pm 2 \ldots$$

The Thermally Perfect Fermi–Dirac Electron Gas

This method of periodic boundary conditions does not change the physics of the problem to any significant extent. We can readily show that the allowed translational energy levels have twice the separation than in the case of fixed-boundary conditions and that the degeneracy of each level is doubled. However, the total number of translational energy states $\Gamma(\varepsilon)$ of equation (2.11) and the

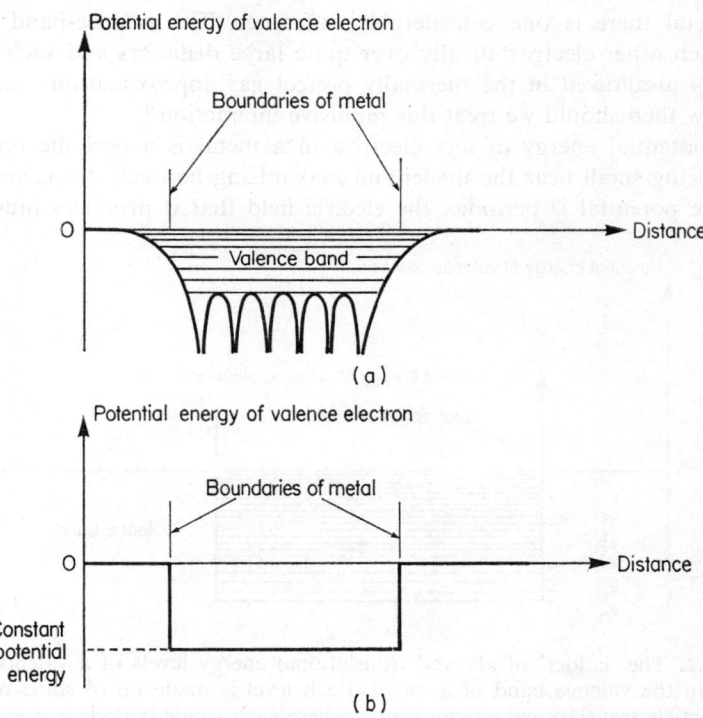

Figure 4.1(a). The actual distribution of potential energy of an electron as a function of its distance from the boundary of the metal. The potential energy rises and falls periodically throughout the volume of the metal. The valence electron is completely free to move in the valence band of translational energy levels. **(b)** The simplified distribution of potential energy of an electron as a function of its distance from the boundary of the metal. The potential energy is constant throughout the entire volume of the metal and forms a potential well with infinitely steep sides.

density of states function, the number of energy states between ε and $\varepsilon + d\varepsilon$ viz. $g(\varepsilon)d\varepsilon$ of equation (2.14) and also the Fermi energy (which we shall introduce in equation (4.1)) are all independent of whether the boundary conditions are fixed or periodic. The reader may wonder why we bother to alter the boundary conditions at all. One advantage is that the periodic boundary conditions eliminate the need for a surface to the metal and also the travelling wave solutions are more closely related to the idea of electrons moving in preferred directions such as in the transport processes considered in Chapter 6.

There is another point to consider which at first sight appears significant. Although we are noting the similarity between the translational energy states of particles in a laboratory container (as in Chapter 2) with the valence electrons in a metal there is one considerable difference. The valence-band electrons repel each other electrostatically over quite large distances and such an interaction is disallowed in the thermally perfect gas approximation—see Section 2.1. How then should we treat this repulsive interaction?

The potential energy of any electron in a metal is a periodic function in space, being small near the nuclei and maximizing between them. Since, however, the potential is periodic, the electric field that it produces must have a

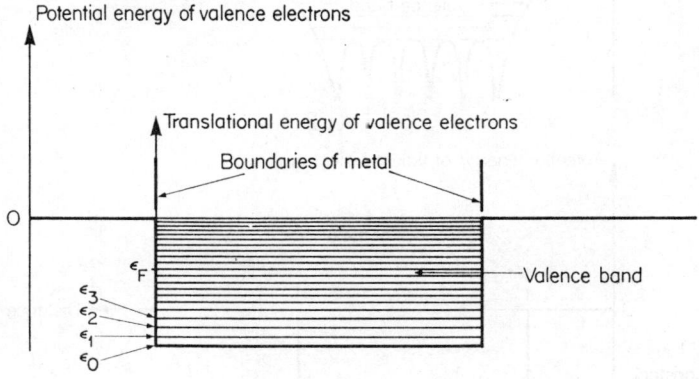

Figure 4.2. The 'ladder' of allowed translational energy levels of a valence electron moving in the valence band of a metal. Each level is made up of some number of single particle translational energy states, where each single particle energy state can be occupied by up to one electron. The energy levels shown in this figure are very similar to those shown in Figure 2.2(a).

value of zero when averaged over many nuclei since the electric field is the gradient of the potential. Figure 4.1(a) shows such a periodic potential. The simplified model that we shall use in this Chapter makes the very drastic assumption that the electrostatic field acting on all of the valence electrons is not only zero when averaged over many nuclei—but is zero *everywhere*. This means the potential energy of all of the valence electrons is a *constant* throughout the entire volume of the metal (see Figure 4.1(b)). Although the potential is constant throughout the metal volume it must increase at the surface of the metal otherwise the valence electrons would escape spontaneously from the metal due entirely to their individual kinetic energies as they move throughout the

The Thermally Perfect Fermi–Dirac Electron Gas

metal, but of course within their valence band. To avoid such a spontaneous loss of electrons we picture valence electrons as moving in a flat-bottomed potential 'well' with infinitely steep sides (see Figure 4.1(b)). Inside this 'well' they have translational, that is kinetic energy but a constant potential energy which we shall define as equal to zero since potential energy is only a relative quantity anyway.

This flat-bottomed potential 'well' with its array of valence band translational energy levels shown in Figure 4.2 is called the valence electron or Sommerfeld free electron model of a metal and its simplicity depends entirely on the fact that the electrons *have only translational*, that is *kinetic energy* which is readily handled, as we have already seen in Chapters 2 and 3. Whether it is a 'good' model, thus whether our drastic assumptions are realistic, depends entirely on how well the theoretical results obtained by it agree with experiment.

Before developing the theory it may help the reader if we give a brief summary of what we have done. We have assumed the valence electrons of all the atoms to be completely free to move throughout the entire volume of the metal such that each valence electron is always in one of the quantized translational energy states comprising the valence band. Each electron has kinetic energy but no potential energy and does not interact electrostatically with any other electron. The electrostatic interaction has already been accounted for in the smoothed out, flat-bottomed potential 'well' and also, of course, indirectly when we applied the Pauli principle, since this arises due to the mutual repulsion of electron wave functions. We can therefore treat the valence electrons as a thermally perfect gas such that the electrons do suffer collisions with each other and also with the boundaries of the metal, but do not interact with each other electrostatically. We are therefore free to use the F.D. statistics of Chapter 2 to determine the thermodynamic properties of this valence-electron gas. In so doing we must remember that every translational energy state in the analysis of Chapter 2 characterized by a quantum number n alone is allowed to have up to *two* electrons in it provided these electrons have different spin quantum numbers.

Historically the thermally perfect valence-electron gas model was analysed before F.D. statistics were known about. These early studies assumed that the valence-electron gas obeyed M.B. statistics and they were able to explain:

a. Ohm's law linking electrical current with electric field.
b. The Wiedemann–Franz–Lorenz law linking electrical conductivity and thermal conductivity.

However, despite these successes, the valence-electron gas of M.B. statistics did not explain:

c. The experimental heat capacity of a high-temperature metal, as this was found to be equal to that of the lattice alone. Stated more succinctly, an insulator was found to have the same heat capacity as a metal, and it looked as if the valence electrons were not participating in the heat-capacity process at all.

We shall demonstrate in this book how F.D. statistics explain all of these points. Let us therefore immediately apply F.D. statistics to the valence-band electrons and see what results we can obtain. In Chapter 2 we showed that the F.D. distribution was given by equation (2.39) and had the form

$$\frac{N_n^*}{g_n} = \frac{1}{\exp(-\alpha)\exp\left(\frac{\varepsilon_n}{kT}\right) + 1}$$

where the energy ε_n referred to a single particle translational energy state that could be occupied by no more than one electron.

We have not yet determined the undetermined multiplier α; let us write

$$\alpha = \frac{\varepsilon_F}{kT} \qquad (4.1)$$

whence,

$$\frac{N_n^*}{g_n} = \frac{1}{\exp\left(\frac{\varepsilon_n - \varepsilon_F}{kT}\right) + 1} \qquad (4.2)$$

This keeps α non-dimensional, provided we treat ε_F as an energy. We can in fact give a physical meaning to this energy ε_F by considering the behaviour of equation (4.2) at $T = 0°K$.

$$\frac{N_n^*}{g_n} = 1 \quad \text{for} \quad \varepsilon_n \leqslant \varepsilon_F$$
$$\frac{N_n^*}{g_n} = 0 \quad \text{for} \quad \varepsilon_n > \varepsilon_F \qquad (4.3)$$

This follows since for those single-particle translational energy states with $\varepsilon_n < \varepsilon_F$, where $(\varepsilon_n - \varepsilon_F)$ is a negative number, the exponential in the denominator of equation (4.2) is the exponential of a very large negative number and is therefore close to zero so that the left-hand side of the equation tends to unity. However, for single-particle translational energy states with $\varepsilon_n > \varepsilon_F$, where $(\varepsilon_n - \varepsilon_F)$ is positive, we have the exponential of a large positive number which is of course a very large number and therefore the left-hand side tends to zero.

The result, that at the absolute zero

$$N_n^* = g_n \quad \text{for} \quad \varepsilon_n \leqslant \varepsilon_F$$

The Thermally Perfect Fermi–Dirac Electron Gas

means that all single-particle translational energy states having energy below ε_F are filled with their Pauli quota of electrons. Figure 4.3 shows a plot of equation (4.2) for two temperatures. We see quite clearly ε_F is the maximum translational energy of the valence electrons at the absolute zero. ε_F is called the Fermi energy or more colloquially the Fermi level. We can now see very clearly the dramatic effect of the Pauli principle on the thermally perfect valence-electron gas. At the absolute zero only two electrons can be in the zero $n = 0$

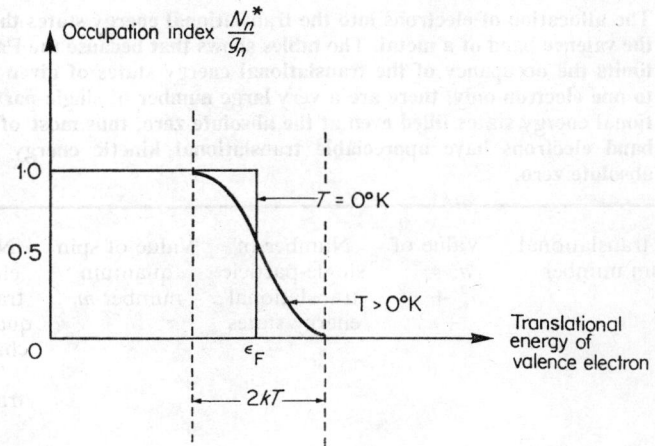

Figure 4.3. Occupation index $\dfrac{N_n^*}{g_n}$ which is the average number of valence electrons per single particle translational energy state as a function of translational energy for the Fermi–Dirac distribution of valence electrons in a metal. Note that at all non-zero temperatures the lower single particle translational energy states have an occupation index of unity (which is their Pauli quota) which falls off to zero in the energy band $2kT$ about the Fermi energy. At the absolute zero, all single particle translational energy states are occupied by one electron (their Pauli quota) right up to the Fermi energy ε_F.

translational energy state and all of the remaining valence electrons must have non-zero kinetic energy where these kinetic energies range all the way up to the Fermi energy.

We can use Table 4.1 to illustrate the manner in which these electrons are packed into a spread of energies even at the absolute zero.

The zero translational energy level ($n^2 = 0$) contains the first two electrons of opposite spins. The third electron must go into the next translational energy level ($n^2 = 1$) which can accommodate 12 electrons. This makes a total of $2 + 12$, 14 electrons allocated. The fifteenth electron goes into the next translational energy level ($n^2 = 2$) which can accept 24 electrons in all totalling $2 + 12 + 24 = 38$ electrons allocated. Proceeding thus we gradually ascend the translational energy levels of the valence band adding two electrons into each translational energy state characterized by n_x, n_y, n_z. This filling process continues until all of the valence electrons have been allocated each to a given single-particle translational energy state. Clearly when all the electrons have been so allocated we shall find that all translational energy levels below a certain energy level (the Fermi level) will be occupied by electrons and all

Table 4.1 The allocation of electrons into the translational energy states that constitute the valence band of a metal. The tables shows that because the Pauli principle limits the occupancy of the translational energy states of given n, m_s, value to one electron only, there are a very large number of single-particle translational energy states filled even at the absolute zero, thus most of the valence band electrons have appreciable translational kinetic energy even at the absolute zero.

Value of translational quantum number			Value of $n^2 =$ $n_x^2 + n_y^2$ $+ n_z^2$	Number of single-particle translational energy states	Value of spin quantum number m_s	Number of electrons in translational quantum level characterized by translational quantum number n
n_x	n_y	n_z				
0	0	0	0	1	$+\tfrac{1}{2}$	2
0	0	0		1	$-\tfrac{1}{2}$	
1	0	0	1	1	$+\tfrac{1}{2}$	12
1	0	0		1	$-\tfrac{1}{2}$	
-1	0	0		1	$+\tfrac{1}{2}$	
-1	0	0		1	$-\tfrac{1}{2}$	
0	1	0		1	$+\tfrac{1}{2}$	
0	1	0		1	$-\tfrac{1}{2}$	
0	-1	0		1	$+\tfrac{1}{2}$	
0	-1	0		1	$-\tfrac{1}{2}$	
0	0	1		1	$+\tfrac{1}{2}$	
0	0	1		1	$-\tfrac{1}{2}$	
0	0	-1		1	$+\tfrac{1}{2}$	
0	0	-1		1	$-\tfrac{1}{2}$	
1	1	0	2	1	$+\tfrac{1}{2}$	
			etc.			

The Thermally Perfect Fermi–Dirac Electron Gas

energy levels above the Fermi level will be empty. Hence the Fermi level *divides the totally filled levels from the empty levels at the absolute zero*.

As the temperature rises from absolute zero we note from Figure 4.3 that only the most energetic electrons are affected. There is an energy band centred about the Fermi energy and of width about $2kT$ within which the number of electrons per single-particle translational energy state changes rapidly from one to zero. We also see that in this case of non-zero temperature the Fermi energy does *not* equal the maximum electron kinetic energy, thus it is quite possible for some electrons to have kinetic energies greater than the Fermi energy level and furthermore the higher the temperature, the greater will be the number of valence electrons with kinetic energies in excess of the Fermi energy. Although we shall apply the Fermi energy concept a great deal in this chapter we should make it quite clear to the reader that the Fermi energy has only a *physical* significance at the absolute zero; for temperatures above absolute zero the Fermi energy, although mathematically a useful quantity, is only a reference energy. This idea of a reference energy follows since we can write equation (4.2) as

$$1 - \frac{N_n^*}{g_n} = \frac{1}{\exp\left(-\dfrac{\varepsilon_n - \varepsilon_\mathrm{F}}{kT}\right) + 1} \tag{4.4}$$

This shows that the distribution is symmetrical with change of sign about the Fermi energy and also about $\dfrac{N_n^*}{g_n} = 0{\cdot}5$ assuming the Fermi energy is independent of temperature. This result is not quite true at very high temperatures but the symmetrical approximation is accurate and very useful, as we shall see in Chapter 7.

Let us return to a remark made above that only those highest energy electrons near the Fermi energy change their translational energy states when the temperature is raised above absolute zero. We shall later discuss electrical and thermal conduction processes in metals and semiconductors in some detail. It is sufficient to remark here that both processes involve electrons gaining kinetic energy either from an electric field or an external heat source if the electrons are to participate in the conduction process. Whether a valence electron can gain kinetic energy is decided by the availability of an empty translational energy state at the new electron energy. If there is an empty translational energy state at the new translational energy then the electron will move into it, otherwise not. The important point is that since electrical and thermal conduction processes involve the valence electrons gaining only *small* amounts of additional kinetic energy, it will only be those valence electrons near the Fermi energy that are permitted to gain this additional kinetic energy. Those valence electrons lying considerably below the Fermi energy level are unable to participate since there are no empty translational energy states close enough for them to enter; put colloquially, the vast bulk of the valence electrons lying 'deeply below' the Fermi energy level do not participate in conduction processes; they lie so deep that they cannot be 'dug out' to participate! It is only those valence electrons that lie close to the Fermi energy that are able to be raised into higher translational energy states and are therefore available to participate in electrical and thermal conduction processes; we shall call them the *'conduction' electrons*. This means that although there are a very large number of valence electrons in a metal (about 10^{28} per cubic metre), only a relatively small fraction of these are available to actually participate in electrical and thermal processes within the metal (we shall find, in fact, what percentage are available in Section 4.5 below). These conduction electrons that do participate lie in the high-energy 'tail' of the translational energy distribution. But we have shown in Figure 2.8 that the high-translational energy 'tails' of all three statistics are essentially the same, thus we can treat the high-energy tail of an F.D. distribution as an M.B. distribution. We therefore get the rather curious result that in spite of the fact that the bulk of the valence electrons in a metal obey F.D. statistics, the most important ones for technological purposes are the conduction electrons, or those in the high-energy 'tail' and they obey M.B. statistics. This is a very important conclusion to have reached and we shall return to it later in Chapters 6 and 7.

The reader may be interested to know under what conditions *all* of the valence electrons rather than just those near the Fermi energy level, would obey M.B. statistics. We can readily answer this question by considering the degeneracy parameter Λ introduced in Section 2.13. In equation (2.75) of that chapter we showed that for copper at room temperature the degeneracy parameter had the value of $1 \cdot 3 \times 10^{-4}$. If we regard the inequality $\Lambda \gg 1$ to be satisfied when Λ has the value 50, say, then we would need to heat copper up to a temperature of about 14,000,000°K before it would behave as an M.B. gas. Since copper melts at 1,356°K our copper would not be a metal, it would be a highly ionized plasma and we have seen in Section 3.3 that most laboratory plasma do indeed obey M.B. statistics.

4.3 The Fermi level

We have now shown that at the absolute zero of temperature the valence electrons are packed up to a translational energy level called the Fermi level. Let us obtain a value of the Fermi energy by using the analysis given in Chapter 2. We shall treat the thermally perfect valence-electron gas as identical to that gas model studied in Section 2.3 except that in this case each translational energy state of translational quantum number n can hold up to two electrons

The Thermally Perfect Fermi–Dirac Electron Gas

provided that they have opposite spins. Figure 4.2 shows the allowed translational energy levels of the electrons sitting in their flat-bottomed potential well and should be compared with Figure 2.2(a).

In Section 2.4 we showed that the translational energy states could be represented by points in the positive octant of a sphere drawn in 'quantum number space'. This sphere was described by equation (2.9) as

$$n^2 = n_x^2 + n_y^2 + n_z^2 = \frac{8m_e V^{\frac{2}{3}} \cdot \varepsilon}{h^2}$$

Here we have put a subscript e on the mass to remind us that we are dealing with electrons. At the absolute zero the sphere has the equation

$$n_{\max}^2 = 8m_e V^{\frac{2}{3}} \varepsilon_{\max} \cdot h^{-2}$$

where $\varepsilon_{\max}$ is the maximum valence electron kinetic energy at the absolute zero, or the Fermi energy. Hence we can write

$$n_{\max}^2 = 8m_e V^{\frac{2}{3}} \varepsilon_F h^{-2} \tag{4.5}$$

We can obtain another expression for $n_{\max}$ since the number of electrons within the octant, N_e will equal *twice* the number of translational energy states in the octant—and the number of such translational energy states equals the volume of the octant. The factor 2 arises since each translational energy state, if occupied, is allowed to contain up to two electrons of opposite spins in it, and of course at the absolute zero every translational energy state is so occupied by its Pauli quota of two electrons.

Therefore
$$N_e = 2 \times \text{Volume of octant}$$

$$= 2 \times \frac{1}{8} \times \frac{4\pi}{3} n_{\max}^3$$

i.e.

$$N_e = \frac{\pi}{3} n_{\max}^3 \tag{4.6}$$

Combining equations (4.5) and (4.6) to eliminate $n_{\max}$ gives

$$\left(\frac{3N_e}{\pi}\right)^{\frac{2}{3}} = 8m_e V^{\frac{2}{3}} \varepsilon_F h^{-2}$$

whence

$$\varepsilon_F = \frac{h^2}{8m_e}\left(\frac{3N_e}{\pi V}\right)^{\frac{2}{3}} \tag{4.7a}$$

$$= \frac{h^2}{8m_e}\left(\frac{3}{\pi} \cdot n_e\right)^{\frac{2}{3}} \tag{4.7b}$$

Here N_e = total number of electrons in the valence band of the metal
n_e = total number of electrons per unit volume of metal, in the valence band (the valence band electron concentration or density).

Let us evaluate equation (4.7) for the case of silver. We require to know the number of valence electrons contributed by each atom of the metal. The alkali metals and the noble metals give 'about 1' valence electron per atom. We cannot be more precise since the valence electron gas model is only an approximation anyway (see Section 4.6). This gives a value of $n_e = 5\cdot 86 \times 10^{28}$ valence electrons per cubic metre (see Appendix 2); on substitution into equation (4.7b) we find a Fermi energy of $9\cdot 0 \times 10^{-19}$ J (5·6 eV). This is the maximum kinetic energy of the valence electrons at absolute zero. The valence-electron gas behaves totally differently from an ordinary Maxwell–Boltmann gas because in this latter case there is *no* thermal energy at all at the absolute zero [see equation (3.3a)]. As regards the metal, it would obviously be completely wrong to regard the absolute zero as a state in which all the valence electron translational motion has ceased. The reader may be puzzled as to why at the absolute zero most of the valence electrons retain kinetic energy, and are moving around when common sense says they should be at rest. Perhaps the valence electrons would come to rest if they were permitted to—but the Pauli principle forbids them to do so and that is that!

4.4 The actual distribution of electrons in the valence band

Equation (4.2) gives the average number of valence electrons in any single-particle translational energy state, viz. $\dfrac{N_n^*}{g_n}$. Let us consider this result for the Fermi distribution

$$\frac{N_n^*}{g_n} = \frac{1}{\exp\left(\dfrac{\varepsilon_n - \varepsilon_F}{kT}\right) + 1}$$

Both sides of this equation give the average number of valence electrons (between 0 and 1) in a particular single-particle translational energy state n. The right-hand side is often called the F.D. 'probability distribution function' $f(\varepsilon_n)$ since it gives the probability that any single-particle translational energy state of energy ε_n is actually occupied by a valence electron.

We have also seen in Sections 2.4 and 2.13 that translational energy levels have an essentially continuous distribution. This is because the energy separation between adjacent translational energy levels is minute and this scale of quantization is so fine we can ignore it. Let us, as we did in Section 2.13, assume that the allowed translational energy levels of the valence band are completely

The Thermally Perfect Fermi–Dirac Electron Gas

continuous and that we can replace the discrete translational energies ε_n by the continuous variable ε—(see the discussion in Section 3.1). In this case the F.D. probability distribution function becomes

$$\frac{N(\varepsilon)\mathrm{d}\varepsilon}{2g(\varepsilon)\mathrm{d}\varepsilon} = f(\varepsilon) = \frac{1}{\exp\left(\dfrac{\varepsilon - \varepsilon_\mathrm{F}}{kT}\right) + 1} \qquad (4.8)$$

Here $f(\varepsilon)$ = probability that a single-particle translational energy state of energy ε contains an electron
= average number of electrons per single-particle translational energy state

Equation (2.14) gives $g(\varepsilon)\mathrm{d}\varepsilon$ which is the number of translational energy states in the energy range $\mathrm{d}\varepsilon$. We must remember that the Pauli principle allows us up to two valence electrons in each of these translational energy states. We have

Total number of valence electrons actually occupying single-particle translational energy states in the range of energy ε to $\varepsilon + \mathrm{d}\varepsilon$ of the valence band = Number of allowed single-particle translational energy states in the energy range × Average number of electrons per single-particle translational energy state

i.e.
$$N(\varepsilon)\mathrm{d}\varepsilon = 2g(\varepsilon)\mathrm{d}\varepsilon \times f(\varepsilon)$$
$$= 2 \times 2\pi(2m_e)^{\frac{3}{2}}Vh^{-3}\varepsilon^{\frac{1}{2}}\mathrm{d}\varepsilon \times \frac{1}{\exp\left(\dfrac{\varepsilon - \varepsilon_\mathrm{F}}{kT}\right) + 1}$$

i.e.
Total number of electrons lying in the energy-state range ε to $\varepsilon + \mathrm{d}\varepsilon$ of the valence band

$$N(\varepsilon)\mathrm{d}\varepsilon = \frac{4\pi V}{h^3}(2m_e)^{\frac{3}{2}} \frac{\varepsilon^{\frac{1}{2}}\mathrm{d}\varepsilon}{\exp\left(\dfrac{\varepsilon - \varepsilon_\mathrm{F}}{kT}\right) + 1} \qquad (4.9)$$

Figure 4.4 shows a schematic plot of equation (4.9) for two different temperatures, one of them being absolute zero. Once again we see that the distribution of low kinetic-energy valence electrons changes very little as the temperature is raised above absolute zero. What happens is that the valence electrons nearest the Fermi energy at zero temperature are raised into translational energy states higher than the Fermi energy for temperatures above absolute zero. This means that:

a. there will be a few valence electrons in states of energy greater than the Fermi energy

b. there will therefore be a few electron 'spaces' left in the states of energy immediately below the Fermi energy.

We can illustrate this idea schematically in Figure 4.5 with Figures 4.2 and 4.4 as a guide. In Figure 4.5 we take a closer examination of those particular translational energy levels near the Fermi level and we ignore the energy levels either considerably below or above the Fermi level. Figure 4.5(a) shows the distribution of electrons in these levels close to the Fermi level at the absolute zero (in practice the energy-level occupation numbers would be very large and

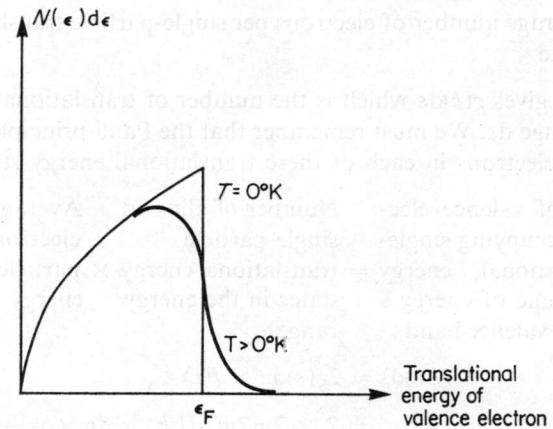

Figure 4.4. Schematic plot of $N(\varepsilon)d\varepsilon$ which is the total number of valence band electrons occupying single particle translational energy states in the range of energy ε to $\varepsilon + d\varepsilon$ versus the translational energy of the single particle translational energy state ε for two temperatures. As in Figure 4.3 the effect of raising the temperature above absolute zero is to thermally excite a narrow band of electrons near the Fermi level into energy states lying higher than the Fermi level, thereby leaving electron energy state vacancies in those energy states immediately below the Fermi level.

not the simple integers shown in the diagram). We see that the level populations numerically increase up to the Fermi energy and then stop abruptly. Figure 4.5(b) shows the same group of energy levels after the temperature of the metal has been raised to a finite non-zero temperature. We see that all the levels below that level in Figure 4.5(a) containing 20 electrons still contain the same number of electrons in Figure 4.5(b). However, the levels of Figure 4.5(a) that originally contained 20, 21, 22, 23 and 24 electrons all lose some electrons to the levels *above* the Fermi level. We therefore see a narrow band of levels on *both* sides of the Fermi level where the level occupancy has altered due to a change in temperature. Clearly the energy states constituting these levels are not filled to their Pauli quota of one electron per single-particle translational energy

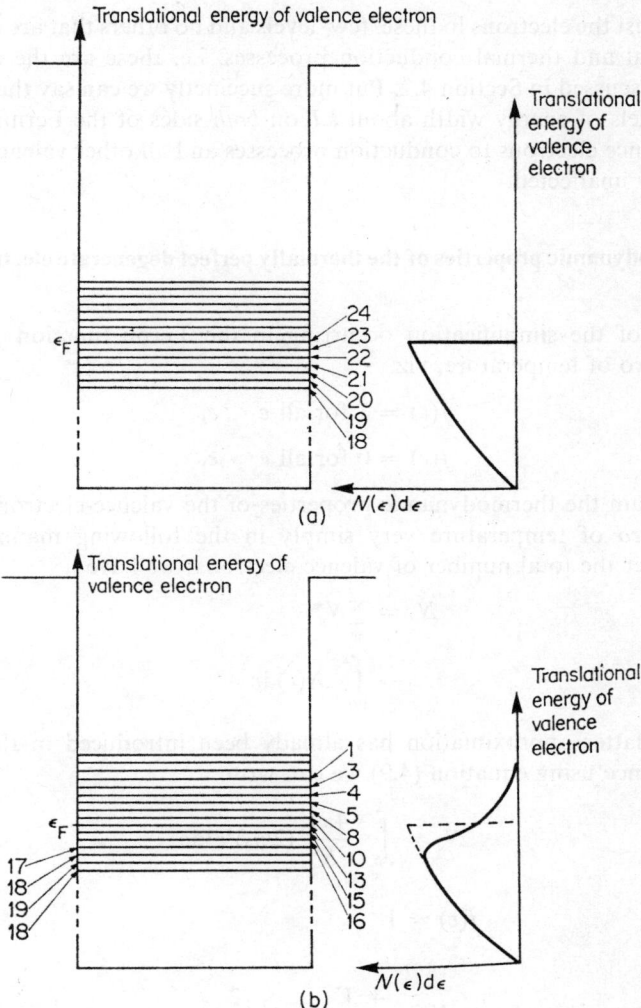

Figure 4.5(a). A magnified version of that group of levels of Figure 4.2 around the Fermi level. The numbers on the levels give the level occupany (number of electrons in a level) at the absolute zero. The graph on the right represents the number of valence band electrons occupying single-particle translational energy states as a function of the translational energy of the valence electrons. (This graph is identical to the $T = 0°K$ graph of Figure 4.4 rotated anti-clockwise through $90°$. **(b)** The same group of levels at a non-zero temperature showing how the level occupancy has changed as valence band electrons originally near the Fermi level have been thermally excited into energy levels higher than the Fermi level. The graph on the right is identical to the $T > 0°K$ graph of Figure 4.4 rotated anticlockwise through $90°$.

state. It is just the electrons in these 'few' levels and no others that are responsible for electrical and thermal conduction processes, i.e. these are the conduction electrons discussed in Section 4.2. Put more succinctly we can say that a narrow band of levels of energy width about kT on *both* sides of the Fermi level contribute valence electrons to conduction processes and all other valence electrons are sensibly unaffected.

4.5 Thermodynamic properties of the thermally perfect degenerate electron gas in a metal

Because of the simplification occurring in the Fermi function $f(\varepsilon)$ at the absolute zero of temperature, viz.

$$f(\varepsilon) = 1 \text{ for all } \varepsilon \leqslant \varepsilon_F$$
$$f(\varepsilon) = 0 \text{ for all } \varepsilon > \varepsilon_F$$

we can obtain the thermodynamic properties of the valence-electron gas *at the absolute zero* of temperature very simply in the following manner. Let us first consider the total number of valence electrons; we have

$$N_e = \sum_n N_n^*$$
$$\to \int_0^{\varepsilon_F} N(\varepsilon) d\varepsilon$$

where the latter approximation has already been introduced in the previous section. Hence using equation (4.9) we can write

$$N_e = \int_0^{\varepsilon_F} \frac{4\pi V}{h^3} (2m_e)^{\frac{3}{2}} \varepsilon^{\frac{1}{2}} d\varepsilon$$

since

$$f(\varepsilon) = 1$$

i.e.

$$N_e = \frac{4\pi V}{h^3} (2m_e)^{\frac{3}{2}} \cdot \frac{2}{3} \varepsilon_F^{\frac{3}{2}}$$
$$= \frac{8\pi V}{3h^3} (2m_e)^{\frac{3}{2}} \varepsilon_F^{\frac{3}{2}} \quad (4.10)$$

This result can of course be rearranged to give

$$\varepsilon_F = \frac{h^2}{8m_e} \left(\frac{3N_e}{\pi V}\right)^{\frac{2}{3}}$$

which is a result already derived in equation (4.7a).

The Thermally Perfect Fermi–Dirac Electron Gas

Now let us consider the total thermodynamic energy of the system, i.e. the valence-electron gas at the absolute zero of temperature. This is

$$E_0 = \sum_n N_n^* \varepsilon_n$$

$$\rightarrow \int_0^{\varepsilon_F} \varepsilon N(\varepsilon) d\varepsilon$$

$$= \frac{4\pi V}{h^3} (2m_e)^{\frac{3}{2}} \int_0^{\varepsilon_F} \varepsilon^{\frac{3}{2}} d\varepsilon$$

i.e.
$$E_0 = \tfrac{3}{5} N_e \varepsilon_F \tag{4.11}$$

where the latter follows from equation (4.7a). We can therefore write

Average energy of a valence electron at the absolute zero of temperature

$$= \frac{E_0}{N_e}$$

$$= \tfrac{3}{5} \varepsilon_F \tag{4.12}$$

where we have used equation (4.11). In a Maxwell–Boltzmann gas the average energy of a particle at any temperature T is given by equations (3.3), viz. $\tfrac{3}{2}kT$ which therefore is zero at the absolute zero. Let us calculate the average energy for silver where we have shown in Section 4.3 that the Fermi energy has the value of 9.0×10^{-19} J or 5·6 eV. The average energy of the valence electrons at the absolute zero is 3/5 of this, viz. 5.4×10^{-19} J or 3·36 eV.

We can also calculate the pressure of the valence-electron gas at the absolute zero which we shall call p_0. Perhaps the reader is puzzled by the idea of the electron gas exerting a pressure at the absolute zero. However, since almost all valence electrons possess kinetic energy at the absolute zero they must be colliding with the surfaces of the metal and therefore exerting a pressure there. We can readily calculate this pressure using equation (2.50),

$$p_0 V = \tfrac{2}{3} E_0$$

i.e.

$$p_0 = \frac{2}{3} \frac{E_0}{V}$$

$$= \frac{2}{3V} (\tfrac{3}{5} N_e \varepsilon_F)$$

$$= \frac{2}{5} \cdot \frac{N_e}{V} \cdot \varepsilon_F \tag{4.13}$$

However, from equation (4.7a) we know that the Fermi energy varies as $n_e^{\frac{2}{3}}$ hence p_0 varies as the 5/3 power of the valence-electron density. Let us calculate this pressure for silver. Using the values given above for the Fermi energy and the valence-electron density we readily obtain

$$p_0 = 2\cdot 2 \times 10^{10} \text{N/m}^2$$

which is $3\cdot 1 \times 10^6$ psia or $2\cdot 1 \times 10^5$ atmospheres. This is an enormous pressure exerted inside the metal by the valence-electron gas at the absolute zero. Why is this pressure not observed outside the metal? The answer is that it is balanced inside the metal by the electrostatic attraction between the valence electrons and their positive parent nuclei which form the lattice.

As we have remarked, the thermodynamic properties of the valence electron gas are readily obtainable at the absolute zero because of the very simple form of the F.D. distribution function, viz. $f(\varepsilon) = 1$. However, for temperatures above absolute zero (and these are of more interest to the electrical engineer and physicist), we must find the thermodynamic properties of the valence-electron gas by using the partition function as a mathematical generating function as described in Section 2.12. For an F.D. gas, equation (2.52) shows that

$$Z_{\text{FD}} = \sum_n g_n \ln\left[1 + \exp\left(\frac{\varepsilon_F - \varepsilon_n}{kT}\right)\right]$$

Once again we shall make the accurate approximation introduced in Section 2.4 and applied in Section 2.13, and will regard the translational levels as continuously distributed; consequently our partition function changes from a summation into an integral

$$Z_{\text{FD}} = \int_0^\infty 2 \ln\left[1 + \exp\left(\frac{\varepsilon_F - \varepsilon}{kT}\right)\right] dg \qquad (4.14)$$

since each energy state is two-fold degenerate because of the two spin-quantum states. Here dg is given by equation (2.14) and thus we get

$$Z_{\text{FD}} = \frac{4\pi}{h^3}(2m_e)^{\frac{3}{2}} V \int_0^\infty \varepsilon^{\frac{1}{2}} \ln\left[1 + \exp\left(\frac{\varepsilon_F - \varepsilon}{kT}\right)\right] d\varepsilon \qquad (4.15)$$

The thermodynamic functions follow by differentiation (see equations 2.54–2.56) and we obtain

$$E = \frac{4\pi}{h^3}(2m_e)^{\frac{3}{2}} V \int_0^\infty \frac{\varepsilon^{\frac{3}{2}} d\varepsilon}{\exp\left(\frac{\varepsilon - \varepsilon_F}{kT}\right) + 1}$$

The Thermally Perfect Fermi–Dirac Electron Gas

and

$$p = \frac{8\pi}{3h^2}(2m_e)^{\frac{3}{2}} \int_0^\infty \frac{\varepsilon^{\frac{3}{2}} d\varepsilon}{\exp\left(\frac{\varepsilon - \varepsilon_F}{kT}\right) + 1}$$

where we have used the rules of differentiation under the definite-integral sign. The integral cannot be evaluated analytically, in closed form, and it is customary to express it as a convergent series. We shall not give the mathematical details of this series further consideration as they add nothing to the thermodynamics! Such an analysis does show that

$$E = E_0 \left[1 + \frac{5\pi^2}{12}\left(\frac{kT}{\varepsilon_F}\right)^2 \cdots \right] \tag{4.16}$$

$$p = p_0 \left[1 + \frac{5\pi^2}{12}\left(\frac{kT}{\varepsilon_F}\right)^2 \cdots \right] \tag{4.17}$$

where E_0 and p_0 are given by equations (4.11) and (4.13). These results might have been expected because we have already shown in Sections 4.2 and 4.4 that only those valence electrons within a very narrow band near ε_F actually contribute to the thermodynamic properties since they alone have empty translational energy states accessible to them; we have already called these special valence electrons the conduction electrons. We might therefore expect our electron gas at a non-zero temperature to behave in only a *slightly* different way from its behaviour at the absolute zero. The above series in equations (4.16) and (4.17) confirms this since the first term is the thermodynamic property at the absolute zero; also each series is a expansion in the variable $\left(\frac{kT}{\varepsilon_F}\right)$. Let us see the size of the variable. We have already shown in Section 4.3 that for silver at the absolute zero the Fermi energy has the value of $9\cdot 0 \times 10^{-19}$ J or $5\cdot 6$ eV. Hence

$$\frac{kT}{\varepsilon_F} = \frac{T}{6\cdot 5 \times 10^4} \tag{4.18}$$

which of course is always considerably less than unity even up to the melting point of silver (1234°K).

In some text books in solid state physics it is the practise to introduce the Fermi temperature θ_F where $\theta_F = \frac{\varepsilon_F}{k}$. For the alkali metals, values of the Fermi temperature range from say 20,000°K to 80,000°K. The reader should be warned that this can be very misleading. The Fermi temperature so defined is not a temperature in a thermodynamic sense; it does not mean that the valence-electron gas is very hot and able to transfer thermal energy in the form of 'heat' to another system. The electron gas is indeed energetic, as we saw in Section 4.3, but it cannot transfer its energy at the absolute zero and, thermodynamically speaking, the entire solid, including electron gas and lattice, is at the absolute zero and if anything, is only in a position to *receive* a heat transfer rather than the other way round!

These thermodynamic properties are of interest because they tell us something about conditions within a metal. Historically one of the most important applications of F.D. statistics was to resolve a problem remarked on in Section 4.2 regarding the heat capacity of a metal as being equal to that of an insulator, which suggests that the valence electrons do not participate in the thermal energy gain when the metal is heated (we shall give a more detailed description of the differences between a metal and an insulator in Chapter 7). We can now explain this apparent anomaly because the valence electrons in a metal are highly degenerate and obey F.D. statistics—and not M.B. statistics, as was assumed by early workers studying the theoretical properties of metals. Equations (4.16) and (4.12) show that

$$E = \frac{3}{5} N_e \varepsilon_F + \frac{\pi^2}{4} N_e \frac{(kT)^2}{\varepsilon_F} \ldots$$

The heat capacity at constant volume of the valence-electron gas is

$$C_V^{\text{electrons}} = \left(\frac{\partial E}{\partial T}\right)_V$$

$$= \frac{\pi^2}{2} k N_e \left(\frac{kT}{\varepsilon_F}\right) \ldots \quad (4.19)$$

This result should be compared with the value of the constant-volume heat capacity of a thermally perfect gas of structureless M.B. particles given in equation (3.6), viz.

$$(C_V)_{\text{MB}} = \tfrac{3}{2} kN$$

We can form the ratio of these heat capacities, obtaining

$$\frac{C_V^{\text{electrons}}}{(C_V)_{\text{MB}}} = \frac{\frac{\pi^2}{2} k N_e \left(\frac{kT}{\varepsilon_F}\right)}{\frac{3}{2} kN} = \frac{\pi^2}{3} \frac{N_e}{N} \left(\frac{kT}{\varepsilon_F}\right)$$

Let us apply this to the case of silver, for example, where $N_e = N$ (monovalent). In view of the fact that we have just shown that

$$\frac{kT}{\varepsilon_F} = \frac{T}{6 \cdot 5 \times 10^4}$$

The Thermally Perfect Fermi–Dirac Electron Gas

we obtain a value of the heat-capacity ratio

$$\frac{C_V^{\text{electrons}}}{(C_V)_{\text{MB}}} = \frac{\pi^2}{3}\left(\frac{T}{6.5 \times 10^4}\right) = \frac{T}{2 \times 10^4} \tag{4.20}$$

Hence the constant-volume heat capacity of a valence-electron gas is only about 1·5% of an ordinary laboratory M.B. gas at room temperature and increases to no more than 5% at 1,000°K. It is therefore not surprising that at ordinary temperatures the valence-electron contribution to the heat capacity of a metal is hardly noticeable and that there is no essential difference between metals and insulators in this respect. Of course the reader must realise that the heat capacity of a metal or insulator also contains a (dominant) contribution from the lattice, and we shall examine this point in Chapter 5.

Let us summarize our conclusions regarding the electrons participating in heat-capacity processes. Consider the effect of a heat transfer to a metal at the absolute zero of temperature. If all of the valence electrons of the metal behaved like the thermally perfect gas of M.B. particles that we considered in Chapter 3, then *all* of the valence electrons would participate in the heat-capacity process, thus each valence electron would have its kinetic energy increased. However, since our valence electrons behave as a thermally perfect gas of F.D. particles then all of the translational energy states are filled up to the Fermi level and only those valence electrons close to the Fermi level, i.e. the conduction electrons can be thermally excited to higher energy levels. Exactly the same reasoning holds if the metal is at a temperature above absolute zero except that because of the 'loosening' of the highest energy valence electrons near the Fermi level there are more empty, 'available' electron translational energy states, thus more valence electrons can participate in the heat-capacity process.

We see that the valence-electron heat-capacity behaviour is entirely dictated by the Pauli principle: if the principle did not hold, all valence electrons would participate, but since the principle must hold, only a 'few' valence electrons participate (the conduction electrons).

We can readily obtain a rough estimate of just how many of the valence electrons are conduction electrons. Let us apply equations (2.11) and (2.14) to form the ratio

$$\frac{\text{number of translational energy states in the interval } \varepsilon \text{ to } \varepsilon + d\varepsilon}{\text{total number of energy states between 0 and } \varepsilon} = \frac{g(\varepsilon)d\varepsilon}{\Gamma(\varepsilon)}$$

$$= \frac{2\pi(2m_e)^{\frac{3}{2}}Vh^{-3}\varepsilon^{\frac{1}{2}}d\varepsilon}{\frac{4\pi}{3}(2m_e)^{\frac{3}{2}}Vh^{-3}\varepsilon^{\frac{3}{2}}}$$

$$= \frac{3}{2}\frac{d\varepsilon}{\varepsilon}$$

Now we have said that only those valence electrons within about kT of the Fermi energy contribute so that we can write the above ratio as $\dfrac{3}{2}\dfrac{kT}{\varepsilon_F}$ Hence, the number of conduction electrons participating in the heat-capacity process is

$$\frac{3}{2}\frac{kT}{\varepsilon_F} \times N_e$$

and for silver this has the value

$$\frac{TN_e}{4\cdot 3 \times 10^4}$$

thus at room temperature about 1·5% of the total number of valence electrons are conduction electrons—a conclusion we have already reached earlier.

We have already stated that the heat-capacity process involves the raising of valence electrons to empty translational energy states; we call this process thermal excitation. The reader may wonder what the physical mechanism of thermal excitation is for those few (relatively speaking) but very important conduction electrons close to the Fermi level. The answer will become apparent in Chapter 5; however, we can remark here that the metal initially takes up the 'heat' transferred to it as thermal energy which it stores in the lattice (phonon) system. We shall show that the conduction electrons interact with the phonons every 10^{-14} seconds, and as a consequence gain energy from the phonons and in turn are thermally excited to higher translational energy states, as we have just described.

At the risk of labouring the point, we must note that equation (4.19) for the constant-volume heat capacity contains N_e which is the *total* number of valence electrons and not just the number of conduction electrons. This is not inconsistent with our conclusion that only a few per cent of these valence electrons actually participate in the heat-capacity process; this is explained in equation (4.19) which contains N_e multiplied by $\left(\dfrac{kT}{\varepsilon_F}\right)$ where $\left(\dfrac{kT}{\varepsilon_F}\right) \ll 1$. We can therefore regard $\left(\dfrac{kT}{\varepsilon_F}\right)$ as a 'scaling down' factor for N_e such that the product $N_e\left(\dfrac{kT}{\varepsilon_F}\right)$ gives (approximately) the number of valence electrons actually participating in the heat-capacity process; thus $N_e\left(\dfrac{kT}{\varepsilon_F}\right)$ gives the (approximate) number of conduction electrons.

4.6 Limitations of the thermally perfect degenerate electron gas (Sommerfeld model)

We have seen that the valence-electron gas model satisfactorily resolves one of the major difficulties of thermal solid-state theory, the contribution of the valence electrons to the heat capacity of a metal at ordinary temperatures.

In fact, this model also gives 'reasonable' results for the paramagnetism and diamagnetism of free electrons, field ('cold') emission of electrons from a surface, the contact potential between two metals, the photoelectric effect and the changes of metallic work function due to surface-adsorbed atoms. However, in many cases, the results are only 'reasonable' and not exact. Furthermore, *all* solids have electrons and yet why is it that under some conditions a proportion of these electrons (the valence electrons) act as if they are free to move throughout the solid as a whole (as in a metal) and yet apparently do not do so in an insulator. Furthermore, the valence-electron gas theory cannot explain how solids which are insulators at low temperatures become semiconductors at higher temperatures; or how the total number of valence electrons in a semiconductor varies with temperature and yet is constant with temperature in a metal.

We shall return to these difficulties in Chapter 7 when we discuss the band theory of solids.

Problems for Chapter 4

1. Apply the Fermi–Dirac probability function to the case of iron to determine:
   a. The probability that a single-particle translational energy state is occupied by a valence electron if the energy state lies 0·1 eV and 1·0 eV above the Fermi energy of 11·5 eV if the iron is at a temperature of 300°K.
   b. The temperature to which the iron must be raised in order that the probability of occupancy by a valence electron of a single-particle translational energy state that lies 0·5 eV above the Fermi energy level is 0·01.
   c. Calculate the separation in energy between two single-particle translational energy states whose probability of occupancy by a valence electron is 0·1 and 0·9 respectively, expressing your answer in units of kT.

   Use the numerical values of the universal constants given in Appendix 1.

   Answers: 1 in 49, 1 in 10^{17}; 1260°K; 4·4.

2. Show that the molar heat capacity at constant volume of the thermally perfect Fermi–Dirac valence-electron gas in aluminium is only 2·2% of the lattice contribution at 300°K given that aluminium has a valency of 3, and an atomic weight of 26·98 kg (kg mole)$^{-1}$, a density of $2·70 \times 10^3$ kg/m^3 and a Debye characteristic temperature of 385°K. You may assume that the lattice contribution to the molar heat capacity at constant volume is $3R$.

CHAPTER FIVE

The Thermally Perfect Bose–Einstein Phonon Gas in a Solid

5.1 The thermodynamic properties of a solid

Historically speaking one of the most interesting aspects of solids was their heat capacity, or the way they 'absorbed' thermal energy when heated. Thermodynamically speaking, any thermodynamic system (of which the solid is just one example), has an infinite number of heat capacities C since heat is not a thermodynamic property—it is an interaction across the system boundary and the amount of heat transferred entirely depends upon the way the system responds to the heat transfer (see Section 1.2). In practise, however, only two heat capacities are of importance, constant volume heat capacity C_V when the system volume does not change during the heat-transfer interaction and constant-pressure heat capacity C_p where the pressure on and within the system does not change during heat-transfer interaction. If we restrict our discussion to that class of thermodynamic systems known as solids, we know that solids generally expand on heating and contract on cooling, thus there will be a volume change. It therefore appears that the constant-volume heat capacity is not relevant here. Furthermore, since laboratory experiments generally take place in the atmosphere, which has a constant ('atmospheric') pressure we see that the constant-pressure heat capacity appears to be the more useful thermodynamic property to describe our experiment. This conclusion is quite true but perhaps a little inconvenient for us. The reason is that the heat-capacity at constant volume is given by

$$C_V = \left(\frac{\partial E}{\partial T}\right)_V \qquad (5.1)$$

and our statistical thermodynamic analyses have usually enabled us to find E either directly from equation (2.3) as we did when we discussed the case of the valence-electron gas at absolute zero [see equation (4.11)] or usually we were required to use the partition function method where the energy is given by the appropriate differential of the partition function as shown in equation (2.54) for B.E. and F.D. particles and equation (2.60) for M.B. particles.

The heat capacity at constant pressure is

$$C_p = \left(\frac{\partial H}{\partial T}\right)_p$$

where H is the enthalpy $E + pV$
i.e.

$$C_p = \left[\frac{\partial(E + pV)}{\partial T}\right]_p$$

$$= \left(\frac{\partial E}{\partial T}\right)_p + p\left(\frac{\partial V}{\partial T}\right)_p$$

Clearly the right-hand side does not look so easy to determine directly from our partition functions. Therefore for simplicity we shall consider the constant-volume heat capacity of a solid. There is no loss in generality in doing this because the two heat capacities are always related for any thermodynamic system so that if we determine the constant-volume heat capacity theoretically we can always deduce the constant-pressure heat capacity, which, as we have just shown, is the more meaningful one for solids.

Experiments carried out in the last century showed that the variation of the heat capacity at constant volume for one mole of any pure solid gave the following behaviour (see Appendix 2 for definition of a mole).

a. At elevated temperatures the constant-volume molar heat capacity is constant and independent of the temperature and numerically equal to $3kN_{AV}$, i.e. $3R$ per kg mole where R is the Universal Gas Constant and where N_{AV} is the Avogadro number which is the number of atoms in one kg mole, $6\cdot02 \times 10^{26}$ atoms per kg mole or kilomol. This is the law of Dulong and Petit and applies to temperatures as high as the melting point of the solid.

b. At cryogenic temperatures the constant-volume molar heat capacity falls below $3R$ and in fact falls towards zero following a T^3 behaviour if the solid is an insulator (which we define in Chapter 7) or following a simple first power of T behaviour if the solid is a metal.

In Chapter 4 we have discussed some thermodynamic aspects of metals and showed that in the free valence electron approximation we could separate out in our analysis the valence electrons from their parent nuclei and treat these valence electrons quite independently. Conversely we can therefore treat the parent nuclei independently of the valence electrons; we shall use this assumption in this chapter. In particular we shall consider some thermodynamic properties of these nuclei which, unlike their valence electron 'offspring' are certainly *not* free to wander at will throughout the volume of the solid. Rather, these nuclei are bound together by very strong interatomic forces and they remain essentially in fixed positions forming the lattice (the 'backbone') of the solid.

A simple but rough picture of any solid lattice, whether it be metal, semiconductor or insulator is to think of a three-dimensional array of masses connected by springs, where each mass is a nucleus and each spring simulates the interatomic force between the nuclei. Each mass (nucleus) has an equilibrium position within this array. At the absolute zero of temperature each nucleus will be essentially at rest (here we shall ignore without explanation the phenomenon of zero-point energy of vibration). Suppose that one portion of the solid is heated; then those nuclei immediately adjacent to that region where the 'heat' is being transferred will start to undergo oscillations about their equilibrium positions. Since one portion of the solid is connected to all other portions through the springs, this oscillatory motion will be transmitted throughout the entire volume of the solid provided the oscillations are not damped out. This collective motion is called the thermal motion of the lattice.

Such a model of a solid was first analysed using quantum theory by Einstein. He ignored the undoubtedly complicated collective oscillatory motion of the nuclei by merely assuming that each nucleus oscillated independently of its neighbours. Before we discuss the Einstein and Debye lattice models it is appropriate to say something about quantum oscillators since these underlie the concept of those very important 'particles' of zero-rest mass called photons and phonons.

5.2 The quantum oscillator

The quantum oscillator appears throughout elementary quantum theory and in fact the entire subject of quantum theory derived from a thermodynamic study of oscillatory systems. Let us think of an oscillator (not necessarily quantized) as something capable of vibrating at some frequency v, i.e. it exhibits a periodic behaviour with time. If the reader likes to have a mechanical picture of an oscillator then he can think of the motion of a mass connected by a spring to an immovable body—or of the motion of two particles joined by a spring when they oscillate along their line of centres. In fact, if these two particles are of opposite electrical signs, then we have the oscillating dipole of electromagnetic theory and if they are both electrically neutral then we have the vibrating diatomic molecule.

It is well known that when any thermodynamic system such as a solid is heated, then it emits electromagnetic radiation in the form of electromagnetic waves radiated into the surroundings of the system. Suppose we take a solid

The Thermally Perfect Bose–Einstein Phonon Gas

and hollow out a cavity inside it. Then if the solid is at any temperature above absolute zero it will be emitting electromagnetic waves into the surroundings outside the body and also into the hollow cavity inside the body. An examination of this cavity radiation will show it to possess all frequencies, each possessing electromagnetic energy such that if we add up all the energies of all the frequencies we obtain the 'thermal' energy of the electromagnetic wave radiation field. It must be noted that these electromagnetic waves originate within the walls of the hollow cavity, from the solid, because these walls are continually emitting electromagnetic waves of every possible frequency and direction into the cavity at a rate critically dependent on the temperature of the solid. When the solid is in thermal equilibrium, the amount of electromagnetic radiation (thermal energy) in the cavity is constant because a dynamic equilibrium exists between the energy emitted by the walls into the cavity and that absorbed by the walls due to radiation from within the cavity. Analysis and experiment show that when this equilibrium between emitted and absorbed electromagnetic waves is reached, the distribution of electromagnetic energy amongst the spectrum of wave frequencies in the hollow cavity does not depend on the shape of the cavity nor the material from which the solid is made but only upon the temperature of the solid.

Electrical engineers and physicists would be interested in determining how much energy there is in each and every electromagnetic wave of a given frequency. A Fourier analysis of the radiation field enables us to find this distribution of energies and shows:

a. the energy associated with each wave of a given frequency has the same mathematical form as the energy of an oscillator whose oscillations are simple harmonic.
b. the spectrum of electromagnetic wave frequencies are such that different waves do not interact, interfere or scatter with each other. We obtain a spectrum of standing waves in the hollow cavity where each standing wave has a certain frequency and consists of two travelling waves in opposite directions, as the standing waves on an electrical transmission line. There is a spectrum of standing waves of all frequencies and as we have just noted, each and every standing wave of a particular frequency does not interact with the standing wave of any other frequency.

This then has introduced the harmonic oscillator. Many physical entities can be represented by harmonic oscillators and in fact the spatial motion of *any* thermodynamic system in stable mechanical equilibrium can be represented near the equilibrium position by simple harmonic motion. We shall not analyse the harmonic oscillator in such generality—let us analyse the behaviour of a harmonic oscillator by discussing one mechanical model of it—the motion of a particle that is constrained to move along an axis such that the particle is tied to some fixed point by 'an agency' such as a spring where this 'agency' exerts a restoring force on the particle proportional to the displacement of the particle from the fixed point. Figure 5.1 shows a picture of such an oscillator. Since the motion is to be harmonic, the spring must obey Hooke's law, which states that

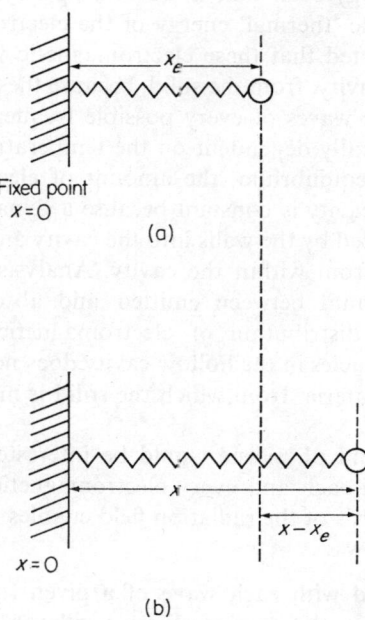

Figure 5.1(a). The mechanical harmonic oscillator when the particle is at rest; the spring connecting the particle to the fixed point is neither stretched nor compressed and has an equilibrium length x_e. **(b)** The mechanical harmonic oscillator in motion. The spring is shown stretched to a total length x so that the extension of the spring from its equilibrium length is $x - x_e$.

the force the spring exerts on the particle is proportional to the extension of the spring which is the amount that the spring is stretched from its unstretched or uncompressed, that is, natural length x_e. Therefore,

$$\text{Force exerted by spring on particle} = F_x$$

where

$$F_x = -f(x - x_e) \tag{5.2}$$

Here f = spring constant, and obviously the restoring force per unit extension of spring. Of course F_x always acts to restore the particle to its equilibrium position x_e and the spring to its natural length because when $x > x_e$ the spring is

The Thermally Perfect Bose–Einstein Phonon Gas

stretched and F_x acts in the negative x direction, i.e. towards $x = 0$ from x positive whilst if $x < x_e$ then the spring is compressed and F_x acts in the positive x direction; this is why there is a minus sign in equation (5.2). Clearly the particle will oscillate about the equilibrium point with simple harmonic oscillations. Using the well-known mechanics formula

$$\text{Potential energy of a stretched spring} = 0.5 \times \text{Tension} \times \text{Extension}$$
$$= \tfrac{1}{2} \cdot F_x \cdot (x - x_e)$$
$$= \frac{f}{2}(x - x_e)^2 \tag{5.3}$$

we have that at any extension

Total energy of oscillator = kinetic energy of oscillator + potential energy of oscillator

$$\varepsilon = \tfrac{1}{2}\mu v^2 + \tfrac{1}{2} f(x - x_e)^2 \tag{5.4}$$

where μ = mass of the oscillator. We now apply Newton's second law, viz.

$$\mu \frac{d^2 x}{dt^2} = F_x$$

i.e.

$$\mu \frac{d^2 x}{dt^2} = -f(x - x_e)$$

The general solution of this equation is

$$x - x_e = A \cos\left[\sqrt{\frac{f}{\mu}} \cdot t + \phi\right]$$

where A = amplitude of the oscillation
ϕ = initial phase of the oscillation

Let

$$\sqrt{\frac{f}{\mu}} \cdot t + \phi = \theta$$

$\therefore$

$$x - x_e = A \cos \theta \tag{5.5}$$

But we have the identity $\cos \theta = \cos(2n\pi + \theta)$ which of course shows that the motion is periodic, or oscillatory, whenever $\sqrt{\dfrac{f}{\mu}} \cdot t = 2\pi$. Let τ = time required for the motion to repeat itself; then

$$\sqrt{\frac{f}{\mu}} \cdot \tau = 2\pi$$

But, frequency of oscillator $= \nu$
$$= \frac{1}{\tau}$$
whence
$$\nu = \frac{1}{2\pi}\sqrt{\frac{f}{\mu}} \tag{5.6}$$
so that
$$x - x_e = A\cos(2\pi\nu t + \phi) \tag{5.7}$$

Since the cosine of any angle varies between $+1$ and -1 we see that $(x - x_e)$ varies between $+A$ and $-A$ so that A is indeed the amplitude of oscillation.
Also,
$$\text{velocity of particle at any instant} = \frac{dx}{dt}$$
$$= -2\pi\nu A \sin(2\pi\nu t + \phi)$$

Since the sine of any angle likewise varies between $+1$ and -1 we see that the velocity can only vary between $\pm 2\pi\nu A$ and that the maximum velocity at any point in the period is
$$v_{\max} = 2\pi\nu A$$

If we combine equations (5.4) and (5.7) we obtain
$$\text{Total energy of oscillator} = \frac{1}{2}\mu\left(\frac{dx}{dt}\right)^2 + \frac{1}{2}f(x - x_e)^2$$
$$= \frac{\mu}{2}[2\pi\nu A \sin(2\pi\nu t + \phi)]^2$$
$$+ \frac{1}{2}\cdot(2\pi\nu)^2\cdot\mu[A\cos(2\pi\nu t + \phi)]^2$$
$$= 2\mu(\pi\nu A)^2 \tag{5.8}$$
$$= \text{Constant}$$

In subsequent chapters we shall also require the following results. Average of square of extension over a period of oscillation
$$= \overline{(x - x_e)^2}$$
$$= \frac{\int_0^{\frac{1}{\nu}} (x - x_e) dt}{\int_0^{\frac{1}{\nu}} dt}$$
$$= \frac{A^2}{2} \tag{5.9}$$

Also

Average of potential energy in one oscillation period

$$\overline{\tfrac{1}{2}f(x-x_e)^2}$$

$$= \tfrac{1}{4}fA^2$$

$$= \mu(\pi v A)^2 \qquad (5.10)$$

$$= \tfrac{1}{2}(\text{Energy of oscillator}) \qquad (5.11\text{a})$$

Likewise we can show that
Average of kinetic energy in one oscillation period

$$= \mu(\pi v A)^2$$

$$= \tfrac{1}{2}(\text{Energy of oscillator}) \qquad (5.11\text{b})$$

The above formulae which have all been derived from classical Newtonian mechanics are the classical equations of a harmonic oscillator; clearly there is no notion here of energy quantization.

If we now return to our discussion of the hollow cavity electromagnetic radiation in a solid we can then replace the entire thermal energy electromagnetic radiation field by the total energy of a collection of simple harmonic oscillators. If we use the above classical simple harmonic oscillator equations together with some thermodynamics, we unfortunately get the wrong answer for the total energy distribution versus frequency of the radiation field; this was the problem Planck attempted to solve. He realized that each oscillator is some oscillatory entity that can emit and absorb electromagnetic energy but only energy at the frequency of oscillation of the oscillator itself. In order to fit the theory of a harmonic oscillator into the experimentally observed energy distribution of the radiation field, Planck found it necessary to assume that the energy of a 'radiation' oscillator was restricted to an integral multiple of a basic unit of energy hv, thus the energy of a radiation oscillator of frequency v is

$$\varepsilon_m = mhv \qquad (5.12)$$

where m = oscillator quantum number 0, 1, 2, . . .
hv = the basic unit of radiation oscillator energy.

Such radiation oscillators therefore emit or absorb electromagnetic energy of amounts equal to an integral multiple of hv; hence, if a radiation oscillator has energy $\varepsilon_{m_1}(v)$ before emitting or absorbing energy and has an energy $\varepsilon_{m_2}(v)$ after, then

$$\varepsilon_{m_1}(v) - \varepsilon_{m_2}(v) = |m_1 - m_2|hv \qquad (5.13)$$

We note that the oscillator, although it may gain or lose energy, never changes its frequency of oscillation. The actual process of energy transfer between oscillators of the same frequency (and we must recall that a radiation oscillator can only affect radiation of the same frequency as that of the oscillator itself) involves an intermediate agent—the photon. In the radiation-oscillator energy change described by equation (5.13) the radiation oscillator has emitted or absorbed $|m_1 - m_2|$ photons. We see that a photon is characterized by a frequency v and has energy proportional to this frequency; by definition the energy of a photon is always hv.

Let us summarize what we have done. We have analysed the electromagnetic energy distribution in the hollow cavity of any solid at a given temperature and shown that this thermal electromagnetic energy can be regarded as being produced by a collection of non-interacting simple harmonic radiation oscillators where the radiation oscillator energy obeys *both* the classical and quantum energy equations, viz.

$$\varepsilon = \tfrac{1}{2}\mu v^2 + \tfrac{1}{2}f(x - x_e)^2$$
$$= mhv$$

Of course the second requirement, the requirement of energy quantization means that the amplitude of oscillation and also the maximum velocity are likewise quantized. This follows from equations (5.10) and (5.11), viz.

$$\varepsilon = 2\mu(\pi v A)^2 = mhv$$

whence

$$A = \left(\frac{mh}{2\pi^2 v \mu}\right)^{\tfrac{1}{2}}$$

Also

$$v_{\max} = \pm 2\pi v A = \pm \sqrt{\frac{2mhv}{\mu}}$$

As far as our purposes in this book are concerned, it is best for us to think of a situation such that each radiation oscillator at any time possesses an energy mhv equal to some integral number m of a basic energy unit hv and also where each radiation oscillator can emit or absorb photons. Hence the electromagnetic (and also thermodynamic) properties of this thermal electromagnetic radiation field can be obtained by studying the behaviour of this collection of radiation oscillators. We can view such a study somewhat differently and bring it into the framework already introduced in our discussion of the thermally perfect gas. Any electromagnetic standing wave of frequency v has energy mhv and therefore is equivalent to and replaceable by a radiation oscillator of energy mhv, which in turn is equivalent to m photons each of energy hv. Since

The Thermally Perfect Bose–Einstein Phonon Gas

there is a spectrum of frequencies in the hollow cavity, we picture this hollow cavity as being filled with photons, each photon carrying energy $h\nu$, there being many photons for each frequency ν and an unlimited number of frequencies.

Development of this approach gives the thermally perfect photon gas in the hollow cavity. Since there can be many photons of frequency ν the photon gas obeys BE statistics. We shall not analyse the cavity photon gas since it gives the Planck black-body radiation law which is of little immediate interest to undergraduate electrical engineering and physics students. Rather, we shall turn our attention from the hollow cavity to the *walls* of the cavity and consider what processes are occurring within the walls. The walls of this cavity constitute a solid lattice composed of atoms which, as we have stated above, can be regarded as a three-dimensional array of masses (nuclei) connected by springs. The analysis of such a mechanical network can be a difficult problem. The first quantal analysis was given by Einstein, who ignored the complexities of the coupling between the oscillating nuclei and assumed that each oscillating nucleus (which we shall call a 'material' oscillator) oscillated completely independently of all its neighbouring nuclei and that each nucleus (material oscillator) performed simple harmonic oscillations about its equilibrium position. Furthermore, all nuclei were assumed to oscillate with the same frequency unlike the radiation oscillators which had a complete spectrum of frequencies. Surprisingly enough, this astonishing collection of simplifications gave answers to the thermodynamic problems of the high-temperature behaviour of a solid lattice that agreed closely with experiment. However, it did fail to predict the correct behaviour of, for example, the heat capacity of the solid as the temperature fell towards absolute zero. We shall not discuss the Einstein theory since it has been superseded by a better theory due to Debye.

5.3 The Debye theory of the lattice

Since the interatomic forces that hold the solid together, preserving the lattice structure, are very strong it would appear reasonable to try to make some allowance for the fact that the motion of each nucleus must surely affect the motion of the other nuclei in the immediate neighbourhood. If we allow for this mutual interaction between nuclei, then a more exact description of the thermal motion of the lattice is one where we think not in terms of the oscillations of the individual nuclei, but rather in terms of the collective oscillations of the entire array of nuclei comprising the lattice. Here our analysis applies to any solid lattice, be it that of a metal, semiconductor or insulator. We shall go to the other extreme of the Einstein model, thus instead of treating the oscillating lattice nuclei as independent quantum oscillators we shall 'de-focus' our picture of the lattice which we know to be made up of discrete nuclei and treat it as a continuum, a continuous piece of matter. This was the model suggested by Debye. He imagined any solid as a continuous elastic medium, with the thermal motion of the lattice not as individual oscillations of the nuclei but as an elastic wave motion moving throughout the solid, the thermal energy of the entire solid (as far as the lattice is concerned) being carried in an array of thermal waves within the solid.

The student of electrical engineering or physics may wonder why it is necessary for him to concern himself with the thermal motion of the lattice of a solid; what concern is it to him? Perhaps the most relevant answer to this is that the electrical conductivity (or resistivity) of a metal or semiconductor is very dependent on the thermal motion of the lattice, as we shall show in subsequent chapters. We must therefore make a simple model of the thermal behaviour of the lattice if we are to understand electrical conduction quantitatively. It is very easy to derive the thermodynamic properties of this model, as we shall do in this chapter and perhaps more importantly apply thermodynamic analysis to introduce the important concept of a phonon.

Returning now to the Debye picture, Debye supposed that the thermal waves were identical to ordinary elastic waves and that elasticity theory could be used to describe the thermal waves. We need not discuss elasticity theory here. All we need to know is that the passage of elastic waves through a solid is controlled by the density and elastic moduli of the solid. There are two types of elastic wave possible; one longitudinal and two transverse. Furthermore, if the solid obeys Hooke's law such that the restoring force on any very small piece of solid is directly proportional to the displacement of that piece from its equilibrium position [see equation (5.2)] then elastic waves will produce a simple harmonic motion of any very small piece of solid and as a consequence the elastic waves travel within the solid with no damping, i.e. without loss of energy. Furthermore, the velocity of propagation of these elastic waves is independent of the wavelength or frequency of the waves. For a solid of finite extent, there will be an infinite number of elastic waves if the solid is truly continuous, each wave being characterized by a different wavelength. Debye could not allow such an infinite number of waves and he had to make some concession to the 'atomicity' of the solid; he assumed that there would be a finite but very large number of elastic waves of different wavelengths allowed. The lower limit to the allowed elastic wavelength or, what is the same thing, the upper limit to the allowed elastic wave frequency, must be dictated by the fact that elastic waves with wavelengths less than the spacing between the lattice nuclei would not propagate at all. Furthermore, the upper limit to the allowed elastic wave wavelengths, or the lower limit to their allowed frequencies, must be dictated by the overall size of the solid since we cannot have a wave whose wavelength is greater than the length of the solid!

These elastic waves move freely throughout the lattice without energy loss from one face of the solid to the other where they will be reflected back into the solid. The elastic waves moving to the right, for example, encounter elastic waves moving to the left and the result is that interference occurs between waves of the same wavelength moving in opposite directions. This results in the formation of a set of standing waves in the solid, each characterized by a certain wavelength or frequency. These standing waves are entirely analogous to the standing waves found on a transmission line due to reflections at the ends of the line. In fact it is the boundary conditions that dictate the array of allowed wavelengths of the elastic waves just as the boundary conditions determined the allowed wave functions and energies of the matter waves which we considered when we solved the problem of the motion of a particle in a box (see Section 2.3). If we therefore write down in order of decreasing wavelength (or increasing frequency) all of the allowed standing waves, we have written down the 'natural' or 'characteristic' frequencies of the elastic waves in our solid. These natural frequencies are all independent of each other and are dictated by the boundary conditions, the dimensions of the solid.

How many natural frequencies are there? In a truly continuous solid there would be an infinite number. However, our solid is really made up of N nuclei bound together. A three-dimensional collection of N particles held together by elastic springs can be shown to have $3N$ natural frequencies; Debye therefore assumed that his solid could also have $3N$ natural frequencies.

A problem now arises. Our assumptions so far treat the thermal motion of the lattice as being carried by an array of standing elastic waves and there are $3N$ such waves, each with a different frequency. How do we bring energy quantization into the argument? Debye assumed that the energy of each and every elastic wave is quantized and this means that there must be a basic unit of energy of an elastic wave. This basic unit of quantized energy of an elastic wave is called a phonon and by definition a phonon has energy $h\nu$ where ν is the frequency of the elastic wave. Put succinctly, the quantum of energy in an elastic wave is a phonon. Let us be quite clear as to what we have done. We have shown that the thermal motion of the lattice is describable in terms of an array of standing waves in the solid. We then replaced each standing wave of this array by a quantum harmonic oscillator such that the oscillator has quantized energy $mh\nu$, i.e.

energy of a standing elastic wave = energy of material oscillator of
of frequency ν frequency ν

$$= mh\nu$$

But $h\nu$ is, by definition, the energy of a phonon of frequency ν; hence, the energy of our standing elastic wave which is equal to the energy of the material oscillator also equals m phonon energies. Clearly there is a strong analogy between photons and phonons. Both arise by considering the energy-versus-frequency distribution of standing waves (electromagnetic waves for photons, thermal elastic waves for phonons) established in a spatial region; both have energy $h\nu$. Photons are the energy carriers of electromagnetic thermal energy in a hollow cavity whilst phonons are the energy carriers of the thermal energy of a lattice. Photons travel at the speed of light, whilst phonons travel at the speed of sound in the medium. Just as we could replace our hollow cavity radiation oscillators by a photon gas, so we can replace our material oscillators by a phonon gas by removing the lattice and replacing it by an equivalent volume of phonons.

5.4 The phonon gas

Since the thermal energy of the solid is equal to the thermal energy carried by all $3N$ elastic waves, we can say that the thermal energy of the solid is found by merely adding up all the phonon energies. We can therefore think of the solid as a volume full of phonons, each phonon possessing energy $h\nu$ and momentum $h\nu a^{-1}$ where a is the speed of propagation of the phonon, which we shall take as the speed of sound in the solid. We must use the word 'momentum' in a non-Newtonian sense since the phonon has zero mass and a flux of phonons does not involve a transfer of mass; in this respect phonons and photons are clearly similar.

Because of the similarity of determining the energy of a collection of non-interacting particles by merely adding up their individual quantized energies [see equation (2.3)], and in the present case the thermal energy of the lattice by adding up the non-interacting phonon energies we can talk of the thermally perfect phonon gas within the solid and use our statistical results of Chapter 2 to describe the phonon gas. There are, however, several differences between a phonon (or photon) and a material particle. The latter is always localized into some definable region of space, subject to the Heisenberg Uncertainty Principle. However, both photons and phonons were introduced as basic energy units of either electromagnetic or elastic thermal waves and a simple wave, although in principle possessing a precise frequency (or wavelength) has clearly no definite position since it can be extended over the entire volume accessible to it. If such waves are to be compared to the behaviour of particles of non-zero rest mass, we must be able to localize them so that the region of localization is where the 'particle' is known or expected to be. In normal wave theory, if we wish to obtain a wave so restricted to a definite region of space then we must form a wave packet consisting of a group of waves of slightly different frequencies but whose phases and amplitudes are so chosen that they interfere constructively over a small region of space where localization is required and destructively outside this small region of space. Clearly such a

The Thermally Perfect Bose–Einstein Phonon Gas

wave packet, although localized, will have a certain 'fuzziness' in its spa a boundaries and we cannot assess an exact wavelength or frequency to the wave packet. Wavepackets move with the group velocity.

Hence, in our Debye model, when we picture the allowed standing thermal elastic wave motion as a phonon gas, we are considering any single phonon as being associated with the whole solid. This does not really matter when we are concerned with the *thermodynamic* properties of the phonon gas, because the localization of the phonons does not seem to arise directly in the analysis. However, when we come to analyse transport processes in a solid, as we shall do in the next chapter, we find that we need to consider thermodynamic systems whose thermodynamic properties vary throughout the volume of the system and that we are concerned with the scattering collisions between phonons in localized regions of the system. Clearly, we cannot do this if every phonon is spread throughout the entire volume of the solid; hence, to permit a discussion of phonon-scattering processes, we must form phonon wave packets in the manner just described and we shall imply phonon wave packets when we use the word 'phonon' henceforward in this book.

Returning to our analogy between material particles and phonons we also note that phonons carry energy and exert a pressure at the inner surfaces of the solid because they can be thought of as exchanging momentum at these inner surfaces. It would therefore appear reasonable to treat the phonon gas as a thermally perfect gas since it obeys all the requirements stated in Section 2.1 where the thermally perfect gas was defined. However, there is another important distinction between an 'ordinary' thermally perfect gas made up of material particles (which have a non-zero rest mass) and a thermally perfect phonon gas consisting of zero rest-mass phonons. This difference is that the total number of phonons in the solid is not constant, thus *equation (2.2) is not obeyed*. The process of transferring heat to the solid results in more and more phonons appearing within the solid and in fact we shall, in the analysis below, calculate the relationship between the number of phonons and the temperature of the solid. Phonons can split up by one phonon being spontaneously divided into two and conversely two phonons can spontaneously unite to form a single phonon; in this respect phonons and photons show similar behaviour.

Let us see how we can make a macrostate description of the phonon gas analogously to the macrostate description of a thermally perfect gas of material particles discussed in Section 2.2. Consider the $3N$ different elastic waves (different in frequency). We know that each elastic wave has an energy equal to an integral multiple of phonon energies where each phonon energy associated with any particular elastic wave is different from the phonon energy associated with any other elastic wave since it has a different frequency. Therefore we can write that each standing elastic wave consists of N_n phonons of frequency ν_n where each phonon has energy $h\nu_n$ so that the elastic wave has a total energy

$N_n h\nu_n$ and there are $3N$ elastic waves of different frequency. Hence we might expect a macrostate description of the phonon gas to be,

N_0 phonons of frequency ν_0 and each of energy $\varepsilon(\nu_0) = h\nu_0$
N_1 phonons of frequency ν_1 and each of energy $\varepsilon(\nu_1) = h\nu_1$
etc.
N_n phonons of frequency ν_n and each of energy $\varepsilon(\nu_n) = h\nu_n$
etc.

However this macrostate description is subject to two considerations, firstly, there is no exclusion principle for phonons and we shall treat them as bosons, i.e. there can be more than one phonon in a phonon quantum state. We can therefore apply equation (2.39) appropriate to B.E. statistics

$$\frac{N_n^*}{g_n} = \frac{1}{\exp(-\alpha)\exp\left(\frac{\varepsilon_n}{kT}\right) - 1}$$

How do we interpret this equation for phonons. As we have already remarked, the number of phonons in the gas is not constant, which is completely opposite to the behaviour of the thermally perfect gas of material particles of non-zero rest mass. In our statistical analysis of that gas we considered a closed container with N particles where N was a constant [see equation (2.2)]. However, for the thermally perfect phonon gas N is not a constant—it can vary and equation (2.35) does not apply. The process of transferring heat to a solid results in more and more phonons appearing within the solid. Hence the equation above must have $\alpha = 0$ and we obtain

$$\frac{N_n^*}{g_n} = \frac{1}{\exp\left(\frac{\varepsilon_n}{kT}\right) - 1}$$

Secondly, the number of phonon quantum states is not infinite (as it was for the translational energy states of a thermally perfect gas of material non-zero rest-mass particles), thus n does not go from zero to infinity; it goes from zero to $3N$. We can therefore say that there are a total of $3N$ quantum states such that there are N_n phonons of frequency ν_n in each of these states. Our macrostate description is therefore, more precisely,

N_0 phonons of frequency ν_0 and each of energy $\varepsilon(\nu_0) = h\nu_0$
N_1 phonons of frequency ν_1 and each of energy $\varepsilon(\nu_1) = h\nu_1$
etc.
N_{3N} phonons of frequency ν_{3N} and each of energy $\varepsilon(\nu_{3N}) = h\nu_{3N}$

The Thermally Perfect Bose–Einstein Phonon Gas

Following accepted custom, let us replace the symbol ν_{3N} by the symbol ν_{max}. In this macrostate description a phonon of frequency ν_n has an energy $h\nu_n$. We note that although we know the number ($3N$) of phonon quantum states (or the number of allowed frequencies), we do not know the actual numerical values of the frequencies themselves, the ν_n values. In particular, we do not know the value of ν_{max}, which is the most important parameter in Debye theory, as we shall see.

Analogously to Section 4.4 we can say that the B.E. distribution becomes, bearing in mind that $\alpha = 0$,

$$f[\varepsilon(\nu_n)] = \frac{1}{\exp\left[\dfrac{\varepsilon(\nu_n)}{kT}\right] - 1} \tag{5.14}$$

where $f[\varepsilon(\nu_n)]$ is the phonon probability distribution function, the probable number of phonons having an energy $\varepsilon_n = h\nu_n$. We shall apply this result in Section 5.7.

Let us now return to the calculation of thermodynamic properties of a phonon gas. The partition function for our phonon gas is given by equation (2.52) with $\alpha = 0$, i.e.

$$Z = -\sum_{n=0}^{3N} g_n(\nu_n) \ln\left[1 - \exp\left\{-\frac{\varepsilon(\nu_n)}{kT}\right\}\right] \tag{5.15}$$

As we have seen in the Debye model there are very many allowed frequencies ν_n ranging from $\nu_0 \sim 0$ to ν_{max}, a total of $3N$ frequencies in all. As N is a very large number, about 10^{28} particles per cubic metre, we may accurately replace the summation by an integral

$$Z \to -\int_0^{3N} \ln\left[1 - \exp\left\{-\frac{\varepsilon(\nu)}{kT}\right\}\right] dg(\nu) \tag{5.16}$$

This is entirely equivalent to the procedures used in Chapters 2, 3 and 4 where we showed that if the scale of quantization was 'very fine', we could effectively ignore it and replace the array of allowed quantized energies by a continuous distribution of energies. In the present case we have replaced our array of natural frequencies by a continuous distribution of frequencies with an upper limit, or, viewed alternatively, we have replaced our array of phonon quantum states by a continuous array of such states, once again there being an upper limit to the number of such states.

We cannot integrate equation (5.16) until we have an expression for $dg(\nu)$ which can be interpreted as either the number of natural frequencies lying between frequencies ν and $\nu + d\nu$ or the number of phonon quantum states lying between energies $h\nu$ and $h(\nu + d\nu)$. The theory of elasticity shows that if the velocity of the longitudinal elastic wave is a_L and that of the two transverse waves a_T then

$$dg(\nu) = 4\pi V \left(\frac{1}{a_L^3} + \frac{2}{a_T^3}\right) \nu^2 d\nu \qquad (5.17)$$

where both velocities are assumed to be independent of frequency. Since there are $3N$ natural frequencies in the solid we can integrate equation (5.17) and obtain

$$\int_0^{\nu_{\max}} 4\pi V \left(\frac{1}{a_L^3} + \frac{2}{a_T^3}\right) \nu^2 d\nu = 3N \qquad (5.18)$$

where $\nu_{\max}$ is the upper limit to the allowed frequencies, there being $3N$ of them.

Hence,

$$\frac{4\pi V}{3} \left(\frac{1}{a_L^3} + \frac{2}{a_T^3}\right) \nu_{\max}^3 = 3N \qquad (5.19)$$

whence on substituting this result back into equation (5.17) we obtain

$$dg(\nu) = 9N \frac{\nu^2 d\nu}{\nu_{\max}^3} \qquad (5.20)$$

Equation (5.16) then becomes

$$Z = -\int_0^{3N} \ln\left[1 - \exp\left\{-\frac{\varepsilon(\nu)}{kT}\right\}\right] 9N \frac{\nu^2 d\nu}{\nu_{\max}^3}$$

$$= -\frac{9N}{x_{\max}^3} \int_0^{x_{\max}} x^2 \ln(1 - e^{-x}) dx \qquad (5.21)$$

where

$$x = \frac{h\nu}{kT} \quad \text{and} \quad x_{\max} = \frac{h\nu_{\max}}{kT}$$

Equation (5.21) is the required partition function for our phonon gas and we can use it to determine all of the thermodynamic properties we require using for example equations (2.54) to (2.56), i.e.

$$E = kT^2 \left(\frac{\partial Z}{\partial T}\right)_V = 3kNT \left[\frac{3}{x_{\max}^3} \int_0^{x_{\max}} \frac{x^3 dx}{e^x - 1}\right] \qquad (5.22)$$

The integral cannot be given in terms of common functions and it is usual to define the Debye function as

$$D(x_{\max}) = \frac{3}{x_{\max}^3} \int_0^{x_{\max}} \frac{x^3 dx}{e^x - 1}$$

The Thermally Perfect Bose–Einstein Phonon Gas

We can immediately use equation (5.1) to find the heat capacity. This is

$$C_V = \left(\frac{\partial E}{\partial T}\right)_V = \frac{9kN}{x_{max}^3} \int_0^{x_{max}} \frac{x^4 e^x dx}{(e^x - 1)^2} \qquad (5.23)$$

Here the integral can be integrated by parts to give

$$C_V = 3kN \left[\frac{12}{x_{max}^3} \int_0^{x_{max}} \frac{x^3 dx}{e^x - 1} - \frac{3x_{max}}{e^{x_{max}} - 1}\right] \qquad (5.23a)$$

We see that the first term on the right-hand side is the Debye function and the second is known as an Einstein function. Table 5.1 gives values of these two functions for various values of x_{max}.

The phonon gas pressure is rather awkward to obtain so we shall give its derivation. We have from equation (2.56)

$$p = kT \left(\frac{\partial Z}{\partial V}\right)_T$$

$$= kT \left(\frac{\partial \nu_{max}}{\partial V}\right)_T \left(\frac{\partial Z}{\partial \nu_{max}}\right)_T$$

$$= h \left(\frac{\partial \nu_{max}}{\partial V}\right)_T \left(\frac{\partial Z}{\partial x_{max}}\right)_T$$

$$= h \left(\frac{\partial \nu_{max}}{\partial V}\right)_T \left[-\frac{3Z}{x_{max}} - \frac{9N}{x_{max}} \ln(1 - e^{-x_{max}})\right]$$

where the term in square brackets follows from the rules of differentiation under the integral sign. From equation (5.21), integrating by parts, we obtain

$$Z = -3N \ln(1 - e^{-x_{max}}) + \frac{E}{3kT} \qquad (5.24)$$

using equation (5.22). Hence

$$p = -h \left(\frac{\partial \nu_{max}}{\partial V}\right)_T \left(\frac{E}{kTx_{max}}\right)$$

$$= -\frac{1}{x_{max}} \left(\frac{\partial x_{max}}{\partial V}\right)_T \cdot E$$

$$= -\left[\frac{\partial (\ln x_{max})}{\partial (\ln V)}\right] \frac{E}{V}$$

Table 5.1 Values of the Debye and Einstein functions.

x_{max}	$D(x_{max}) = \dfrac{3}{x_{max}^3} \int_0^{x_{max}} \dfrac{x^3 dx}{e^x - 1}$	$\dfrac{x_{max}}{e^{x_{max}} - 1}$
0	1·000	1·000
0·2	0·927	0·903
0·4	0·858	0·813
0·6	0·793	0·730
0·8	0·732	0·653
1·0	0·674	0·582
1·2	0·621	0·517
1·4	0·571	0·458
1·6	0·524	0·405
1·8	0·481	0·356
2·0	0·441	0·313
2·2	0·404	0·274
2·4	0·370	0·240
2·6	0·339	0·209
2·8	0·310	0·181
3·0	0·284	0·157
3·2	0·259	0·136
3·4	0·237	0·117
3·6	0·217	0·101
3·8	0·199	0·087
4·0	0·182	0·074
4·2	0·166	0·064
4·4	0·152	0·055
4·6	0·140	0·047
4·8	0·128	0·040
5·0	0·118	
5·5	0·095	
6·0	0·078	
6·5	0·064	
7·0	0·053	
7·5	0·044	
8·0	0·037	
8·5	0·031	
9·0	0·026	
9·5	0·022	
10·0	0·019	

The Thermally Perfect Bose–Einstein Phonon Gas

But from equation (5.19)

$$\nu_{\max}^3 = \frac{9N}{4\pi V}\left(\frac{1}{\frac{1}{a_L^3}+\frac{2}{a_T^3}}\right)$$

whence, on taking logarithms and differentiating, we obtain

$$\left[\frac{\partial(\ln x_{\max})}{\partial(\ln V)}\right]_T = -\frac{1}{3}$$

giving the final result

$$p = \frac{1}{3}\frac{E}{V} \tag{5.25}$$

This result should be compared with equation (2.50) and the reader's attention is drawn to the discussion following that equation. This phonon pressure is restrained from disintegrating the solid by the very strong interatomic forces that bind the lattice nuclei together.

5.5 Asymptotic behaviour of the thermodynamic properties of a phonon gas

We have now derived all of the thermodynamic properties that we require in terms of $x_{\max}$ which is in terms of $\nu_{\max}$, which is, as already noted, an important unknown constant of the theory. In fact $\nu_{\max}$ can only be determined by comparison of experiment with theory, as we shall discuss below. From equation (5.19) we can get some idea of the size of $\nu_{\max}$ which has already been shown to be

$$\nu_{\max}^3 = \frac{9N}{4\pi V}\left(\frac{1}{a_L^3}+\frac{2}{a_T^3}\right)^{-1}$$

If we take $\frac{N}{V} \sim 10^{28}$ atoms per cubic metre and the velocity of the elastic waves (the phonon velocity) to be about 5×10^3 metres per second, we find $\nu_{\max}$ has a value between 10^{12} and 10^{13} Hz, which corresponds to a minimum elastic wave wavelength of about 1 to 10 angstrom units. Since the interatomic spacing is about 1 angstrom unit we might expect the continuum approximation made by Debye to be in error at the highest frequencies near $\nu_{\max}$.

It is usual to introduce the Debye temperature $\theta_D = \frac{h\nu_{\max}}{k}$. Typical values are given in Table 5.2; more extensive data are given in Table 5.3.

Let us discuss the temperature dependence of the heat capacity in terms of this reference (Debye) temperature. In fact, we shall find that the Debye temperature provides a reference temperature for not only the heat capacity but also the electrical and thermal conductivities of solids (see Chapter 6).

Table 5.2 Values of the Debye temperature θ_D for various metals, semiconductors and insulators; all of these values have been obtained from experimental heat-capacity studies using Debye theory.

Metal	θ_D
Copper	343°K
Molybdenum	450°K
Lead	110°K
Silver	226°K
Platinum	240°K

Semiconductor	θ_D
Silicon	640°K
Germanium	370°K
Indium antimonide	200°K
Lead telluride	130°K

Insulator	θ_D
Diamond	2230°K

The reader may wonder why we study heat capacities and their temperature dependence. The reason will emerge as the book develops; briefly, a solid is an exceedingly complex quantized system and we can learn something about its energy states from studies of the heat capacity and also from transport processes. Any progress in solid-state theory must come from a combination of accurate experiment and good theory, and this is why there has been an intensive investigation of the thermal aspects of solids. If we wish to examine the temperature dependence of some aspects of a solid we need to cover a wide range of temperature—say a factor of 100. Such a temperature range for a solid cannot be obtained by starting at room temperature and moving up the temperature scale since we would end up by studying plasmas rather than solids! We can only obtain a factor of 100 in the temperature of a solid by starting at room temperature and *descending* the temperature scale towards absolute zero; consequently we must also be concerned with the cryogenic behaviour of solids in this text as well as the behaviour from room temperature up to the melting point of the solid.

The Thermally Perfect Bose–Einstein Phonon Gas

As we have noted, these thermodynamic formulae cannot, in general, be evaluated in closed form. However, we can readily consider their limiting values thus;

High-temperature limit $T > \theta_D$.

Since $x = \dfrac{h\nu}{kT}$ and we let $T \to \infty$ then $x \to 0$ and $e^x \to 1 + x$ therefore

$$E \sim \frac{9kNT}{x_{max}^3} \int_0^{x_{max}} \frac{x^3 dx}{(1+x)-1}$$

$$= 3kNT \qquad (5.26)$$

whence the heat capacity at constant volume is from equation (5.1).

$$C_V = \left(\frac{\partial E}{\partial T}\right)_V = 3kN \qquad (5.27)$$

and for one mole of solid where $N = N_{AV}$ and $kN_{AV} = R$ we obtain

$$\overline{C_V} = 3R \qquad (5.27a)$$

which is the Dulong–Petit value referred to in Section 5.1; hence the Debye theory agrees with experiment at high temperatures where 'high' means considerably greater than θ_D. In fact, equations (5.23) and (5.27) are found to agree within 1% when $T > 2\theta_D$ based purely on arithmetical (as opposed to physical) considerations.

Physically, this means that at these high temperatures there is sufficient thermal energy in the solid to excite all $3N$ elastic waves regardless of their frequency and they are all contributing to the heat capacity.

Also from equation (5.25) we have the equation of state

$$p = \frac{1}{3}\frac{E}{V}$$

$$= \frac{1}{3}\frac{3kNT}{V}$$

i.e.

$$pV = NkT \qquad (5.28)$$

which is identical to equation (3.5). This means that the equation of state of the phonon gas at high temperatures is identical to that of the 'perfect' or 'ideal' gas of mechanical engineering thermodynamics!

Low-temperature limit.

Since $x_{max} = \dfrac{h\nu_{max}}{kT} = \dfrac{\theta_D}{T}$ we can let $x_{max} \to \infty$, where $T > 0$ implies

$x_{max} \to \infty$. The significance of this is that we need place no limit on the maximum frequency ν_{max}. The reason is that no elastic waves having frequencies anywhere near ν_{max} are excited—only the low-frequency elastic waves contribute so that it makes no difference whether we cut the integral off at ν_{max} or infinity; there is essentially no area under the integral curve from ν_{max} to infinity. The fact that only the low frequencies are excited means that the Debye continuum approximation will be very accurate. If we let the upper limit of the definite integral tend to infinity, then the integral becomes the Riemann zeta function and we obtain

$$E = \frac{3\pi^4}{5} \cdot \frac{NkT}{x_{max}^3}$$

$$= \frac{3\pi^4}{5} Nk \frac{T^4}{\theta_D^3} \tag{5.29}$$

We note from this result that the vibrational or thermal energy of the lattice is proportional to T^4 at low temperatures which is analogous to the black-body radiation law (Stefan's law). There is as we have already see, a strong resemblance between the behaviour of phonons and photons which can be fully developed by B.E. statistics. In the latter case the photon gas has no longitudinal mode of propagation—it has two transverse modes (polarization), both of which propagate at the velocity of light.

We also have the heat capacity at constant volume from equation (5.1),

$$C_V = \left(\frac{\partial E}{\partial T}\right)_V$$

$$= \frac{12}{5} \pi^4 Nk \left(\frac{T}{\theta_D}\right)^3 \tag{5.30}$$

which is the cubic temperature dependence already referred to in Section 5.1. It was, in fact, this cubic dependence which was observed experimentally and unexplained by the Einstein discrete model of the lattice that prompted Debye to postulate his continuum model. Equation (5.30) allows a determination of the Debye temperature from experiment since we can measure the constant-volume heat capacity as a function of temperature at low temperatures and select a value of the Debye temperature corresponding best with the experimental data. This method of determining the Debye temperature is known as the calorimetric method. Figure 5.2 shows a non-dimensional form of a heat-capacity curve with typical experimental points. Of course the Debye temperature can be calculated from the elastic properties of the solid or from measured velocities of sound; agreement with calorimetric values of the Debye temperature is quite good.

The Thermally Perfect Bose–Einstein Phonon Gas

Finally, from equation (5.25) we have

$$pV = \tfrac{1}{3}E$$
$$= \frac{\pi^4}{5} Nk \frac{T^4}{\theta_D^3} \qquad (5.31)$$

which is very different from the perfect-gas equation of state that applies at high temperatures [equation (5.28)]. We can get some idea of how 'low' the

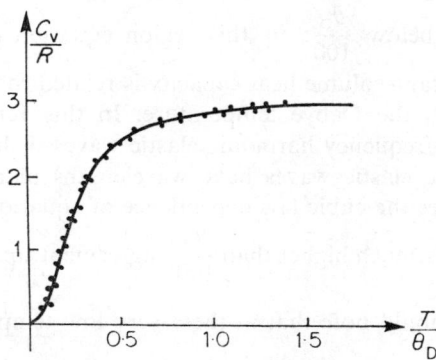

Figure 5.2. Non-dimensional constant volume heat capcity $\frac{C_V}{R}$ of a Debye solid versus non-dimensional temperature $\frac{T}{\theta_D}$. The dots show typical experimental points for a wide range of solids.

temperature must be if we note that equations (5.23) and (5.30) agree arithmetically speaking within 1% for $T < \frac{\theta_D}{12}$ and this agreement becomes better for lower temperatures.

Of course, this discussion of the high and low-temperature behaviour of the constant-volume heat capacity has been in arithmetical terms of the agreement between an exact mathematical expression as given in equation (5.23) and the limiting mathematical approximations as given in equations (5·27) and (5.30). It has been unrelated to the *observed* physical behaviour of solids so that when we come to compare the theory with experiment we may expect to find a somewhat different range of agreement. Experiment shows that Debye theory works satisfactorily for temperatures above about $0·2\theta_D$. In this region equation (5.27) holds, the molar heat capacity at constant volume has the value of $3R$ J (kg mole deg C)$^{-1}$. The significance of this result, as we have already remarked, is that the thermal motion of the lattice 'blurs' the individual features (discreteness) of any particular lattice model and we obtain this very general result which is of course completely independent of the lattice structure since it contains no lattice parameters. What this means is that the nuclei are oscillating essentially independently of each other, (which is the original Einstein assumption) and that the concept of a coherent lattice oscillation is unimportant. All material oscillators are contributing and there are no more oscillators to excite; furthermore the majority of the most energetic oscillators have a frequency of oscillation close to ν_{max} which again was one of the Einstein assumptions.

Experiment also shows that Debye theory works well for the very low temperature region below $\frac{\theta_D}{100}$; in this region equation (5.30) holds which tells us that the constant-volume heat capacity is related to the lattice structure via the parameter θ_D the Debye temperature. In this very low temperature region, only the low-frequency harmonic elastic waves or harmonic oscillators are excited and these elastic waves have wavelengths many times the lattice spacing. In some cases the cubic law dependence of equation (5.30) is found to hold for temperatures much higher than $\frac{\theta_D}{100}$; agreement up to $\frac{\theta_D}{10}$ is sometimes found. The reader should note that at these very low temperatures of $\frac{\theta_D}{100}$ the lattice heat capacity is very small. For example, in the case of copper, the room-temperature value of the constant-volume molar heat capacity is readily calculated from equation (5.27) and found to be about 24×10^3 J(kg mole deg C)$^{-1}$ whilst at 4°K the value is 3.1 J(kg mole deg C)$^{-1}$; the value at 4°K is nearly 10^4 lower than that at room temperature. One immediate consequence of this very low heat capacity at cryogenic temperatures is that thermal equilibrium between solids is very rapidly established.

If we come to study the intermediate temperature range $\frac{\theta_D}{100} < T < \frac{\theta_D}{5}$ then we find that Debye theory does not, in general, hold because the Debye temperature θ_D ceases to be constant and becomes a function of temperature. The reason for this failure is that although equation (5.17) gives $dg(\nu) \propto \nu^2$, in fact in practice $dg(\nu) \propto \nu^p$ where $p > 2$. This means that whilst there is agreement between experiment and Debye theory for low temperatures $T < \frac{\theta_D}{100}$ where the frequencies of the standing waves are small, at higher temperatures where higher standing-wave frequencies (or harmonic oscillator frequencies) are contributing there are more harmonic oscillators actually participating in the heat-capacity process than allowed for in equation (5.17). Thus more thermal energy is required to raise the temperature of the lattice and so the heat capacity increases more rapidly than predicted by Debye theory. Such an increase in the heat capacity corresponds to a decrease in the Debye temperature; the Debye temperature falls and eventually reaches a minimum. However, as the temperature is raised to the region where $T \sim \frac{\theta_D}{5}$ all of the standing waves (or harmonic oscillators) are excited so that further 'addition of heat' does not change the number of participating oscillators and the Debye temperature becomes constant once again.

Table 5.3 gives values of the Debye temperature in the two temperature regions where Debye theory holds.

Table 5.3 Values of the Debye temperature θ_D at 'low' $\left(T < \frac{\theta_D}{100}\right)$ and 'high' $\left(T > \frac{\theta_D}{5}\right)$ temperatures.

Solid	$\theta_D \left(T < \frac{\theta_D}{100}\right)$	$\theta_D \left(T > \frac{\theta_D}{5}\right)$
Na	152	160
K	89	100
Cs	39	46
Cu	343	320
Ag	226	220
Au	162	185
Zn	310	245
Cd	215	165
Al	428	385
Pb	107	88
Bi	119	120
Mo	450	380
W	400	315
Ni	440	390
Pt	240	225

Notwithstanding this temperature dependence of the Debye temperature, it is still a most useful parameter for characterizing the thermal behaviour of the lattice at all temperatures because we can learn a great deal from knowing its numerical value. For example, since $\theta_D = \frac{h\nu_{max}}{k}$ and from equation (5.6) we have $\nu \propto \sqrt{\frac{f}{\mu}}$ for a harmonic oscillator, we see that a high value of θ_D implies that we are dealing with a lattice with very strong interatomic forces (restoring force f) and light atoms; an outstanding example of this is diamond, which has a

Debye temperature of 2230°K. An interesting feature of heat-capacity studies of diamond is that studies made at room temperature are in the 'intermediate temperature' region where Debye theory is known to be inaccurate; this does not hold for any other solid at room temperature.

Clearly the Debye model is an over-simplification. For example, most solids are not isotropic and have in fact different elastic properties in different spatial directions. Also the speed of elastic waves in a solid depends upon the frequency of the wave and also on the direction in which it is travelling. Very much more sophisticated mathematical models of the solid have been examined by finding the natural frequencies of three-dimensional arrays of masses connected by springs. As far as the heat-capacity problem is concerned, these elaborate models give much the same answer as the Debye model; the reason is that the heat capacity at constant volume is not critically dependent on the frequency distribution of the natural frequencies of the solid. We have remarked above that $dg(\nu) \propto \nu^p$ and surprisingly such a simple power law appears to be quite adequate to give good heat-capacity theory. It is clearly of enormous advantage to be able to characterize the heat capacity behaviour of a complex thermodynamic system like a solid by the single parameter θ_D and this is why Debye theory is still widely used. For example, we have already seen that the very large changes (10^4 or more) in the constant-volume molar heat capacity can be represented by the use of a single parameter θ_D. Perhaps the greatest limitation of the Debye model is that it does not allow for the thermal expansion of the solid (we will prove this in Section 6.6) because it does not allow lattice waves (or phonons) of different frequencies to interact with each other (see Section 5.4). This results in the inevitable conclusion that a phonon gas would never regain thermal equilibrium once disturbed from it! We shall return to this difficulty in Chapter 6 when we discuss phonon–phonon interactions and the thermal conductivity of the lattice.

Finally, the Debye model also fails to explain the heat capacity of *metals* at very low temperatures—say within a few degrees of absolute zero. At these very low temperatures the valence electrons in the metal make a significant contribution (see Section 4.5), and this we shall now discuss.

5.6 The total heat capacity of a solid

We have now discussed the two constant-volume heat capacities of a solid, that due to the valence electrons, where equation (4.19) showed that

$$C_V^{\text{electrons}} = \frac{\pi^2}{2} k N_e \left(\frac{kT}{\varepsilon_F} \right)$$

We recall from the final paragraphs of Section 4.5 that $\left(\frac{kT}{\varepsilon_F} \right)$ is a 'scaling down' factor for the total number of valence electrons N_e to give the number of conduction electrons that actually participate. Also the heat capacity due to the lattice is given by equation (5.23a)

$$C_V^{\text{lattice}} = 3kN \left[4D(x_{\max}) - \frac{3x_{\max}}{e^{x_{\max}} - 1} \right]$$

Because of our assumption made in Chapter 4 and also in Section 5.1 that we could completely decouple the valence electrons from the lattice, the total constant-volume heat capacity of the solid is simply the sum of the two contributions.

$$C_V^{\text{total}} = C_V^{\text{electrons}} + C_V^{\text{lattice}} \tag{5.32}$$

In Section 4.5 we concluded that, at room temperature and below, the valence electron contribution to the heat capacity of a metal or semiconductor is exceedingly small and that the heat capacity of any solid, be it metal, semiconductor or insulator, is essentially determined by the heat capacity of the lattice alone. We must show a little caution in making this statement for the following reason. Let us consider temperatures very much less than the Debye temperature. Then we can write

$$C_V^{\text{total}} = \frac{\pi^2}{2} kN_e \left(\frac{kT}{\varepsilon_F} \right) + \frac{12\pi^4}{5} kN \left(\frac{T}{\theta_D} \right)^3 \tag{5.33}$$

where we have used equation (5.30). This can be written as

$$C_V^{\text{total}} = \gamma T + AT^3$$

or

$$\frac{C_V^{\text{total}}}{T} = \gamma + AT^2 \tag{5.34}$$

where

$$\gamma = \frac{\pi^2}{2} \left(\frac{k^2 N_e}{\varepsilon_F} \right) \quad \text{and} \quad A = \frac{12\pi^4}{5} \cdot \frac{kN}{\theta_D^3}$$

Hence, a graphical plot of $\dfrac{C_V^{\text{total}}}{T}$ versus T^2 should give a straight line of slope A and intercept γ. At very low temperatures the term AT^3 approaches zero much faster than the term γT. Hence for metals and semiconductors the valence electron contribution to the constant-volume heat capacity *can* be detected at the very lowest temperatures, since the total heat capacity will fall towards zero as γT, linearly with temperature. Of course above this temperature range the valence electron contribution is swamped by that of the lattice.

This conclusion only holds for metals and semiconductors since they alone have sufficient of their valence electrons actually as conduction electrons to be technologically important. In Chapter 7 we shall show that an insulator possesses effectively no valence electrons as conduction electrons and therefore has no contribution to make to the valence-electron heat capacity. In this case the total heat capacity of the insulator is solely that of the lattice which, as we have just shown, falls to zero as AT^3; we have already referred to this point in Section 5.1 as the distinction in behaviour between conductors and insulators at very low temperatures.

In practise, equation (5.34) does not hold numerically for some metals; a very low-temperature linear behaviour is found but the value of γ does not agree with theory. Let us calculate a theoretical value for copper

$$\gamma = \frac{\pi^2}{2} kN_e \left(\frac{k}{\varepsilon_F}\right)$$

On a molar basis $kN_e \equiv R$ for a metal of unit valency

$$\therefore \qquad \gamma = \frac{\pi^2}{2} R \left(\frac{k}{\varepsilon_F}\right)$$

For copper the Fermi energy has been shown to have the value $1 \cdot 13 \times 10^{-18}$ J whence

$$\gamma = \frac{\pi^2 \times 8 \cdot 31 \times 10^3 \times 1 \cdot 38 \times 10^{-23}}{2 \times 1 \cdot 13 \times 10^{-18}} \text{ J(kg mole)}^{-1}(°K)^{-2}$$

$$= 0 \cdot 501 \text{ J(kg mole)}^{-1}(°K)^{-2}$$

Table 5.4 gives some typical values

Table 5.4 The electronic molar heat-capacity coefficient γ for a selection of metals deduced from very low-temperature heat-capacity studies.

Element	γ_{theory}	$\gamma_{experiment}$
	J(kg mole)$^{-1}$ (°K)$^{-2}$	
Na	1·14	1·38
K	1·69	2·08
Cs	2·36	3·20
Cu	0·50	0·70
Ag	0·65	0·65
Au	0·65	0·73
Al	0·63	1·46
Pb	0·94	3·0
Bi	1·05	0·08
Pt	0·59	6·8

The Thermally Perfect Bose–Einstein Phonon Gas

The discrepancy is found to be quite large for multivalent atoms. Part of this discrepancy can be related to the assumption that the mass of a valence electron in the conductor is equal to the mass of a free electron isolated in space. This is not correct because, due to the electrostatic interaction of the valence electrons with the lattice and electron–phonon scattering collisions, the mass pertinent to our calculations is the 'effective' mass. Calculations can be made of the effective mass but it would involve us in more quantum mechanics than is appropriate for this book; using the effective mass improves the agreement.

5.7 The number of phonons in a solid

In later chapters where we discuss the electrical and thermal conductivity of metals and semiconductors, and also the thermal conductivity of insulators, we shall find that their values are largely dictated by the number of scattering collisions that electrons undergo as they interact with the lattice phonons. Accordingly we shall give a simple derivation of the number of phonons in a solid as a function of temperature.

Equation (5.17) gives

Number of elastic waves whose frequencies lie between ν and $\nu + d\nu$

$$= dg(\nu)$$

$$= \frac{9N\nu^2 d\nu}{\nu_{max}^3}$$

Let $\bar{m}$ = average number of phonons per elastic wave

$$= \frac{1}{\exp\left(\dfrac{h\nu}{kT}\right) - 1} \tag{5.35}$$

This result follows from equation (5.14) since $f[\varepsilon(\nu)]$ is the probable number of phonons having energy $\varepsilon = h\nu$. Hence

Number of phonons whose frequencies lie in the interval ν to $\nu + d\nu$

$$= \bar{m} \cdot dg(\nu)$$

$$= \frac{9N\nu^2 d\nu}{\nu_{max}^3 \left[\exp\left(\dfrac{h\nu}{kT}\right) - 1\right]}$$

We find the total number of phonons in the solid at any temperature T by integration and obtain

$$\text{Total number of phonons} = N_p = \int_0^{\nu_{max}} \frac{9N}{\nu_{max}^3} \cdot \frac{\nu^2 d\nu}{\left[\exp\left(\frac{h\nu}{kT}\right) - 1\right]} \tag{5.36}$$

A slightly different way of obtaining this result which may be more clearly understood is to proceed thus

Number of phonon quantum states whose energy lies between ε and $\varepsilon + d\varepsilon$ where $\varepsilon = h\nu$

$$= dg(\nu)$$
$$= \frac{9N\nu^2 d\nu}{\nu_{max}^3}$$

Let $\bar{m}$ = average number of phonons per phonon quantum state

$$= \frac{dN(\nu)}{dg(\nu)}$$
$$= \frac{1}{\exp\left(\frac{h\nu}{kT}\right) - 1} \tag{5.37}$$

where this result comes directly from the B.E. distribution equation (2.39) with $\alpha = 0$.

Average number of phonons whose frequencies lie between ν and $\nu + d\nu$	=	Number of phonon quantum states with frequencies lying between ν and $\nu + d\nu$	×	Average number of phonons per phonon quantum state

$$= \frac{9N\nu^2 d\nu}{\nu_{max}^3} \cdot \frac{1}{\left[\exp\left(\frac{h\nu}{kT}\right) - 1\right]}$$

whence the same result is obtained.

Consider the integral

$$\int_0^{\nu_{max}} \frac{\nu^2 d\nu}{\exp\left(\frac{h\nu}{kT}\right) - 1}$$

The Thermally Perfect Bose–Einstein Phonon Gas

This cannot be expressed in terms of normal functions as it stands. However, we shall only be interested in its asymptotic limits which are readily found.

High temperature limit

$$\exp \frac{h\nu}{kT} \to 1 + \frac{h\nu}{kT}$$

$$\therefore \int_0^{\nu_{max}} \frac{\nu^2 d\nu}{\left(\frac{h\nu}{kT}\right)} = \frac{kT}{h} \cdot \frac{1}{2} \nu_{max}^3$$

whence the number of phonons is

$$N_p = \frac{9NT}{2\theta_D} \tag{5.38}$$

Low temperature limit.

From Section 5.5 we take $\nu_{max} \to \infty$ in the upper limit to the integral

$$\int_0^\infty \frac{\nu^2 d\nu}{\left[\exp\left(\frac{h\nu}{kT}\right) - 1\right]} = \left(\frac{kT}{h}\right)^3 \int_0^\infty \frac{x^2 dx}{(e^x - 1)}$$

$$= \left(\frac{kT}{h}\right)^3 \zeta(2)$$

where $\zeta(2)$ is the Riemann zeta function. Mathematical tables show that $\zeta(2) = 2\cdot 404$. Hence

$$\text{Number of phonons} = 2\cdot 404 \times 9N \left(\frac{T}{\theta_D}\right)^3$$

$$= 21\cdot 64 N \left(\frac{T}{\theta_D}\right)^3 \tag{5.39}$$

Problems for Chapter 5

1. A great deal can be learned about the structure of a crystal lattice from making purely macroscopic measurements on it. Measurement shows that the density of aluminium at 20°C is $2\cdot 70 \times 10^3$ kg/m^3 and that its atomic weight is 26·98 kg (kg mole)$^{-1}$. Acoustic studies show that to a first rough approximation the longitudinal and transverse elastic-wave velocities are both $3\cdot 30 \times 10^3$ m/s.

   Using these data, calculate the Debye characteristic temperature of aluminium and use Debye theory to calculate the phonon gas pressure within aluminium at a temperature of 96·5°K.

   Answers: 386°K; $1\cdot 46 \times 10^7$ Nm^{-2}.

2. The Dulong–Petit law states that at room temperature and above, the constant-volume molar heat capacity of all solids is constant independent of temperature with a numerical value of 2.49×10^4 J(kg mol deg C)$^{-1}$. Show that at 440°K lead obeys the Dulong–Petit law to within 0·5 per cent but that diamond has a heat capacity only approximately 37% of the Dulong–Petit value.

Comment on the anomalous behaviour of diamond at 440°K in terms of the lattice behaviour. In addition, calculate the temperature to which diamond must be raised in order that it has a heat capacity equal to 95% of the Dulong–Petit value.

You may assume that the Debye temperature for lead is 88°K and that for diamond is 2200°K where this latter numerical value has been chosen for ease of arithmetical calculation.

Answer: 2200°K.

CHAPTER SIX

Transport Properties of Solids

6.1 Transport processes and the equilibrium approximation

We now make a significant departure from the topics discussed in the previous chapters. We have considered the thermodynamic properties of thermally perfect gases in a state of thermodynamic equilibrium with their surroundings, that is, in the evaluation of the equilibrium macrostate there were no heat and/or work transfers occurring to the closed system from the surroundings and the thermodynamic state of the system was fixed, unchanging in time in the manner discussed in Section 1.4. We used the fact that the thermodynamic state was constant when we wrote down equation (2.3) $\sum_n N_n \varepsilon_n = E = $ constant.

The undergraduate is often told that thermodynamics can only analyse thermodynamic systems which are in mechanical and thermal equilibrium with their surroundings, and that thermodynamics usually concerns itself with the initial and final thermodynamic states of the system, i.e. before and after heat and/or work transfers have occurred to it. Sometimes we do concern ourselves with the thermodynamic state of the system as it is actually changing its thermodynamic state but we can only do this when the change of state is thermodynamically reversible and therefore all intermediate states can be described in terms of thermodynamic properties (see Section 1.5). However, there is a somewhat different viewpoint to consider; macroscopic thermodynamics (the study of thermodynamics of matter in bulk without regard to its atomic or molecular structure) has been extended into thermodynamically non-equilibrium situations and there is a relatively new field of study called non-equilibrium thermodynamics or, more accurately, thermodynamics of the steady state. A very simple example of the sort of problem that can be analysed with this new field of thermodynamic study is the flow of thermal energy ('heat conduction') down a conductor which has a temperature gradient—or the diffusion of charged carriers at a p–n junction in a semiconductor. It can be shown that the assumptions underlying the thermodynamics of the steady state are ultimately justified by considerations of basic atomic or molecular processes and whether such processes can be reversed in space and time.

A discussion of the thermodynamics of the steady state and its basic assumptions is not appropriate to an undergraduate textbook. What, for example, can we do to study important non-equilibrium processes such as electrical conduction in a conductor due to an externally applied electric field or thermal conduction in a conductor due to an externally applied temperature gradient? Historically, such non-equilibrium situations have been discussed using kinetic theory, which is a study of the rates of atomic processes. Unfortunately if we look into rigorous kinetic theory we find that, in order to analyse such interesting steady-state non-equilibrium processes, we have to find the particle distribution function. This is a mathematical function that gives the spatial positions and momenta of the particles of the system as a function of time as these spatial positions and momenta change due both to interparticle collisions and externally applied forces such as electric fields or temperature gradients.

It is not too difficult to set up the mathematical equation that gives the time rate of change of the particle distribution function as a function of the externally applied forces and interparticle collisions. However, it is found to be an integrodifferential equation containing both the time derivative and also the time integral of the distribution function. This integrodifferential equation, known as the Boltzmann (transport) equation is very difficult to solve and in fact cannot be solved in general terms. It can only be solved for special cases such as a system being only slightly removed from a state of complete thermodynamic equilibrium and even in this case it involves a great deal of messy algebra which is not appropriate to this book!

It looks, therefore, as if we are unable to discuss 'transport processes' at all where we define a transport process in a system as a movement (ultimately due to interparticle collisions) of some quantity that is conserved in nature where the movement is due to the imposition on the system of the gradient of an intensive thermodynamic or mechanical property of the system. This book seems to be full of drastic assumptions and simplifications, so one more such assumption cannot do any more harm! What we shall do is to analyse transport processes 'in the equilibrium approximation' which merely says that we shall forget all about the formidable Boltzmann (transport) equation that gives the distribution function and instead merely assume that the distribution function in any particular very small volume of the system is the same as if no gradient of an intensive thermodynamic or mechanical property existed across that volume element at all!

Such an assumption allows very great simplifications to be made in the mathematics—as we shall see—but strictly speaking it is quite incorrect. Oddly enough it often gives answers that agree to within a few per cent of the answers that are obtained after a great amount of tedious mathematics from the perturbation solution of the Boltzmann (transport) equation! Since we are primarily interested in approximate answers obtained with as little effort as possible, the 'equilibrium approximation' seems an appropriate method for us to choose!

Transport Properties of Solids

All the transport processes that we shall consider obey the general equation

$$\text{Flux of conserved quantity} = -\begin{pmatrix}\text{Transport}\\\text{coefficient}\end{pmatrix} \times \begin{pmatrix}\text{Gradient of}\\\text{intensive property}\end{pmatrix} \quad (6.1)$$

For example, if the system is an electrical conductor, and if the conserved quantity is electrical charge, and the intensive property is electrical potential V_{pot} then

$$\text{Flux of charge} = -\sigma \frac{dV_{\text{pot}}}{dx}$$

where σ is the electrical conductivity of the system, i.e. σ is the transport coefficient more formally known as the coefficient of electrical conduction. However, flux of charge is current density j (charge crossing unit area in unit time) and also the electrical field E_{F} is given by $E_{\text{F}} = -\dfrac{dV}{dx}$

$$\therefore \quad j = \sigma E_{\text{F}} \quad (6.2)$$

which is Ohm's law and we have used the subscript F for field on the electric field symbol to distinguish it from thermodynamic energy E.

Another example is if the system is a thermal conductor and if the conserved quantity is thermal energy and if the intensive property is temperature T then

$$\text{Flux of thermal energy ('heat' flux)} = -\lambda \frac{dT}{dx} \quad (6.3)$$

Here λ is the transport coefficient known as the coefficient of thermal conduction or, more usually, 'thermal conductivity' for short. Equation (6.3) is Fourier's law of 'heat' conduction.

Yet another example is for any system at all where the conserved quantity is mass and the property possessing a gradient is particle concentration, i.e. number of particles per unit volume n, then

$$\text{Flux of particles} = -D \frac{dn}{dx} \quad (6.4)$$

where D is the transport coefficient known as the diffusion coefficient. Equation (6.4) is Fick's law of diffusion.

The problem that arises in equations (6.2) to (6.4) is to determine the transport coefficients σ, λ or D in terms of known, measurable variables. Let us first consider the electrical conductivity.

6.2 Electrical current conduction in metals

Following our discussion at the end of Section 4.2 consider a metal at some temperature above absolute zero such that there will be some valence electrons (a few per cent) in energy states above the Fermi energy. We have called these important few per cent of valence electrons that actually participate in thermal and electrical conduction processes the conduction electrons.

We have already noted in Section 4.2 that these 'few' but vitally important conduction electrons are governed essentially by M.B. statistics, whilst those valence electrons lying deeper are governed by F.D. statistics. We have also calculated the average *thermal* kinetic energy of a particle in an M.B. gas equation (3.3b) which is

$$\bar{\varepsilon} = \tfrac{3}{2}kT \tag{6.5}$$

In the absence of an externally applied electric field the conduction electrons are in thermal equilibrium with the metal as a whole and move through it with random speeds, where these speeds are in part due to the fact that the temperature of the metal *is* above absolute zero and therefore the conduction electrons *do* possess some thermal energy. Now the average thermal kinetic energy is $\tfrac{1}{2}m_e^*\overline{v^2}$ where $\overline{v^2}$ is the mean square thermal speed. But equation (6.5) also gives the average thermal kinetic energy so that we have

$$\tfrac{1}{2}m_e^*\overline{v^2} = \tfrac{3}{2}kT$$

i.e.

$$\sqrt{\overline{v^2}} = \sqrt{\frac{3kT}{m_e^*}}$$

The quantity $\sqrt{\overline{v^2}}$ in the above equation is the square root of the mean square thermal speed; i.e. it is the root mean square thermal speed which we shall denote by the symbol v_{th}. Hence

$$\text{Root mean square thermal speed} = \overline{v_{\text{th}}} = \sqrt{\overline{v^2}} = \sqrt{\frac{3kT}{m_e^*}} \tag{6.6}$$

where m_e^* is the effective mass of the conduction electron in the metal referred to at the end of Section 5.6. In addition to possessing this thermal kinetic energy, the conduction electrons also possess the Fermi energy (see next Section). Any motion of charged particles constitutes an electric current so that there will be current flow in the direction of electron movement. However since the conduction electron motion is random, the net current flow in any particular direction will be zero.

If we want electrical current to flow in a specified direction, we must make the conduction electrons move predominantly in that direction. It is for this reason, that is, that we are interested in electron flow in a specified direction rather than at random, that we introduced the periodic boundary conditions in Section 4.2, since such boundary conditions give travelling electron waves rather than

Transport Properties of Solids

standing electron waves. We can make the conduction electrons move predominantly in one direction by several methods, two of which are:

a. Apply an electric field to the metal so that the conduction electrons move in this field. This will give current flow due to conduction-electron drift and is obviously related to the externally applied electric field.

b. Produce current flow by diffusion, thus applying the fact that nature always tries to 'smooth out' disturbances. More specifically, if we can create an excess of conduction electrons in one place compared with another linked to it, then the random speeds of the conduction electrons will produce electron flow to smooth out the electron-concentration difference. Clearly this does not require an externally applied field—but does require some means of producing local excess concentrations of conduction electrons.

We shall in fact consider both these aspects of electrical conduction. Let us initially consider the problem of electrical conduction in a metal due to an externally applied electric field. We must explain or justify the following observed features:

a. Metals obey Ohm's law, which states that the current density j is proportional to the applied electric field E_F provided that steady state conditions apply and that the applied electric field is not too large. We have the equation $j = \sigma E_F$ where σ is the electrical conductivity already introduced and which is a finite quantity.

b. The electrical conductivities of metals at room temperature lie between 7×10^5 to $7 \times 10^7 \, \Omega \, \text{m}^{-1}$ or their resistivities ρ lie between $1 \cdot 5 \times 10^{-6}$ to $1 \cdot 5 \times 10^{-8} \, \Omega \text{m}$ where ρ is the reciprocal of σ.

c. The resistivity of many metals increases linearly with temperature, provided the metal is at a temperature above about $0 \cdot 5 \theta_D$.

d. Below about $0 \cdot 25 \theta_D$ the resistivity decreases at a rate faster than linearly with temperature—roughly as T^5 for simple (monovalent) metals.

e. Below about $0 \cdot 05 \theta_D$ the resistivity becomes constant and independent of temperature.

f. At very low temperatures, usually a few degrees above absolute zero, many metals and alloys become superconductors. We shall not, however, discuss superconductivity in this book at all.

g. Metals are good conductors of thermal energy ('heat'). Above a temperature equal to θ_D the ratio of thermal conductivity to electrical conductivity is proportional to the first power of temperature such that the constant of proportionality is approximately the same for all metals.

h. The resistivity of most metals decreases with increasing external pressure applied to the metal.

Consider equation (6.2)

$$j = \sigma E_F$$

Here j is the net rate of passage of conduction electrons across unit area lying perpendicular to the direction of current flow. However, current density can also be expressed in terms of the concentration of electrons multiplied by their charge and their drift velocity, i.e.

$$j = n_e(-e)v_e \tag{6.7}$$

What is the meaning of v_e the bulk velocity of movement of the conduction electrons in the electric field? In a perfect electrical conductor, with a perfectly periodic lattice, the conduction electrons would move as if they were in a vacuum! They would accelerate freely in the electric field and this would result in the electrical conductivity of such a perfect conductor being infinite. We need not attempt to prove this conclusion since the passage of an electron through a perfectly periodic lattice is a problem in wave mechanics and the result is given in most of the standard textbooks on solid-state theory. However, the problem is entirely analogous to the transmission of an electromagnetic wave through a wave filter or along a transmission line with periodically distributed elements. Since the filter or line is perfectly lossless, thus non-resistive, all impedances are pure reactances. Such a filter or line has one or more pass bands and one or more stop bands (this is analogous to the allowed bands and energy gaps in a solid, as we shall see in Chapter 7). In the pass bands, the signal passes through without scattering, reflection or attenuation; in the stop bands there is no transmission at all.

Returning to the electrical conduction problem in a metal, we find that the unimpeded acceleration of the conduction electrons does in fact *not* occur. What we do find is that for moderate electric fields the conduction electrons move with a mean velocity proportional to the externally applied electric field. This 'resistance' to the unimpeded electron motion is entirely due to the fact that the lattice is not perfect, that is, deviations from a perfectly periodic lattice electrostatic potential do occur and we shall shortly consider the effect this has. In terms of our wave filter or transmission-line analogy, these lattice imperfections can be considered as follows. Suppose the filter has either L or C values that vary slightly from element to element (in the solid this might represent a dynamic lattice defect, i.e. departure from perfect periodicity due to the thermal motion of a lattice nucleus) or perhaps a totally different L or C value at one particular position (in the solid this might represent a static lattice defect due to the presence of an impurity atom in the lattice). In the first case there would be a large number of small reflections of the signal and in the latter case a small number of almost complete reflections. The same happens in the solid, namely the conduction electrons are scattered.

Whatever the conduction electrons collide with (scattering mechanisms will be discussed in greater detail later on in this chapter), they lose kinetic energy and

Transport Properties of Solids

change direction as they move more or less in the general direction of the applied electric field between scattering collisions. Analogously to the kinetic theory of 'ordinary' gas particles we refer to the mean free path of a conduction electron l as the average distance travelled by any conduction electron between scattering collisions. In each free path a conduction electron gains kinetic energy from the applied field and on colliding with a vibrating lattice nucleus or with a lattice impurity or defect, the electron gives up some kinetic energy and changes direction. This transfer of kinetic energy (which has been gained by the electron from the electric field) to the lattice results in a rise in the thermal energy content of the lattice, thus the lattice temperature rises. This is known as joule or ohmic heating.

In Sections 6.1 and 6.2 we referred to the 'equilibrium approximation' of a transport process requiring a steady state. Where is the steady state in the conduction electron problem? A steady state will come into existence when the energy gained by any conduction electron in a free path between scattering collisions is just equal to the energy lost by that electron when it does make a scattering collision.

6.3 The collisional relaxation time and electron-scattering processes

There is one point that has been glossed over in our use of equation (6.7) concerning the number of electrons participating in the electrical conduction process. In Sections 4.2, 4.4, 4.5 and 6.2 we have carefully pointed out that not all of the valence electrons that make up the electron gas were in fact able to participate in electrical and thermal processes and in fact only those few per cent of valence electrons near the Fermi energy level (the conduction electrons) could participate since these alone had empty higher energy states to move into.

Yet in equation (6.7) we have used the symbol n_e to denote the *total* number density of valence electrons and not just that small percentage near the Fermi energy. This is a subtle point but there is no inconsistency in our reasoning. When an external field is applied to the metal every one of the valence electrons from translational energy zero to the highest energy near the Fermi level accelerates in this field (we note that the field accelerates the electrons in a direction opposite to that in which the electric field is pointing because the electrons are negatively charged). There is always an empty translational energy state ready to receive *any* valence electron (which is changing its energy due to the field) since an empty energy state will have been created by the simultaneous change of energy by another valence electron; hence *all* of the valence electrons move in the electric field.

Hence the velocity of every valence electron increases with time so that the current likewise grows with time. In the field E_F the force on each electron is $(-e) \cdot E_F$ so that by Newton's second law we have

$$(-e) \cdot E_F = m_e^* \frac{dv_e(t)}{dt} \tag{6.8}$$

Here $v_e(t)$ is the velocity of the electron due to the electric field and this velocity changes with time; it is in *no* way related to the thermal speed of the electron which is constant in time. But on a time basis

$$j(t) = n_e(-e) v_e(t)$$

whence,

$$\frac{dj}{dt} = n_e(-e) \frac{dv_e}{dt}$$

$$= n_e(-e) \frac{(-e) \cdot E_F}{m_e^*}$$

$$= n_e \cdot e^2 \frac{E_F}{m_e^*} \tag{6.9}$$

which shows that the current density grows proportionately with time—and furthermore this current density will grow indefinitely large if the electrons are unrestrained in their motion in the applied electric field.

However, the electron movement cannot continue indefinitely because as we have already noted, the electrons meet 'resistance to motion'. We stated that this 'resistance' would not occur if the lattice were perfectly periodic and in such a case we would obtain infinite electrical conductivity. The fact that the electrical conductivity is not infinite but apparently finite means that all lattices must be imperfect and at this stage we shall merely refer to 'scattering processes' as representing these imperfections. Let us suppose that the resistive force caused by these scattering processes is proportional to the effective electron mass and the electron velocity in the electric field, i.e.

$$\text{resistive force} \propto m_e^* \cdot v_e(t)$$

$$= \frac{1}{\tau} \cdot m_e^* \cdot v_e(t) \tag{6.10}$$

Here $\frac{1}{\tau}$ is a constant of proportionality with the dimensions of inverse time. Since the electron moves in a direction opposite to the field, this resistive force opposing the electron motion must likewise operate in a direction parallel to the field. Hence by Newton's second law, our equation of motion allowing for scattering processes becomes

$$m_e^* \frac{dv_e}{dt} = (-e) \cdot E_F - \frac{m_e^* \cdot v_e}{\tau} \tag{6.11}$$

Transport Properties of Solids

The solution of this differential equation consistent with the initial condition that $v_e = 0$ when $t = 0$ (when the field is switched on) is

$$v_e(t) = \frac{(-e) \cdot E_F \cdot \tau}{m_e^*} \left[1 - \exp\left(-\frac{t}{\tau}\right)\right] \qquad (6.12)$$

This shows that the electron velocity once again increases with time but that after a time greater than τ, the velocity of the electron 'settles down' (does not increase further) to a constant value

$$v_e = \frac{(-e) \cdot E_F \cdot \tau}{m_e^*} \qquad (6.13)$$

which is the drift velocity introduced in equation (6.7).

Also, from equation (6.9) we have

$$\frac{dj}{dt} = n_e(-e)\frac{dv_e(t)}{dt}$$

$$= n_e(-e)\left[\frac{(-e) \cdot E_F}{m_e^*} - \frac{v_e(t)}{\tau}\right]$$

$$= \frac{n_e(-e)v_e(t)}{\tau} + \frac{n_e \cdot e^2 E_F}{m_e^*}$$

$$= \frac{j(t)}{\tau} + \frac{n_e \cdot e^2 E_F}{m_e^*} \qquad (6.14)$$

This gives the growth of current density with time and the solution to this differential equation is

$$j = \frac{n_e \cdot e^2 \cdot \tau \cdot E_F}{m_e^*} \left[1 - \exp\left(-\frac{t}{\tau}\right)\right] \qquad (6.15)$$

We note that the asymptotic behaviour of this equation, the value of the current density for times greatly exceeding τ is

$$j = \frac{n_e \cdot e^2 \tau \cdot E_F}{m_e^*} \qquad (6.16)$$

$$= \sigma E_F$$

using equation (6.2). Hence we obtain

$$\sigma = \frac{n_e \cdot e^2 \cdot \tau}{m_e^*} \qquad (6.17)$$

We see that the electrical conductivity of a metal is proportional to the *total* number density of valence electrons and we have shown in Chapter 4 how to calculate this number. In passing it is pertinent to note that this number of valence electrons per unit volume is a constant for a given metal and independent of temperature whereas for a semiconductor, as we shall see in Chapter 7, the conduction-electron concentration is critically dependent on the temperature. Equation (6.17) shows that the electrical conductivity increases with the value of τ. But what is τ? We have already remarked that the entire valence-electron distribution moves when the electric field is first applied but that this movement does not increase indefinitely because the electrons meet a 'resistance to motion', thus the picture is one where the entire electron distribution moves without hindrance for a time τ and then the resistive force is felt. τ is called a collisional relaxation time and is the average time between scattering collisions of the conduction electrons.

To calculate the electrical conductivity using equation (6.17) we therefore must be able to calculate the collisional relaxation time. However, if we return to the valence-electron gas model of Chapter 4 we recall that this model completely ignores the lattice altogether. Now, we do know that electron–electron collisions occur in the thermally perfect electron gas so we might be tempted to calculate a value of the collisional relaxation time for such electron–electron collisions. We shall not give such a calculation because if we make it and employ the result in equation (6.17) then we obtain calculated values of electrical conductivity far larger than experimentally measured values. Furthermore, electron–electron collisions could not provide the necessary 'resistance to motion' because such collisions do not destroy electron momentum in the direction of the electric field. Failing this, we might be tempted to modify the valence-electron gas to allow for some slight electron binding, thus the valence electrons are not completely free, but this would be incorrect. After all, metals do contain free-valence-electrons, that is, unbound ones, because Ohm's law holds right down to vanishingly small voltages. If the valence electrons were even very weakly bound then Ohm's law would only hold above a certain threshold voltage, and this is not the case. We therefore must ascribe the 'resistance to motion' as due to conduction electrons being scattered by lattice imperfections. We can phrase this conclusion succinctly by saying that 'electrical resistance at and around room temperature is solely due to lattice imperfections'.

How can we obtain a numerical value of the collisional relaxation time due to such lattice imperfections? The first step is to obtain a numerical value of the collisional relaxation time by using equation (6.17) with a *known* value of electrical conductivity. To do this we must allow n_e to be the total number density of valence electrons notwithstanding the fact that the relaxation time refers to only those few per cent of valence electrons near the Fermi energy level that actually suffer scattering collisions viz. those conduction electrons lying within a narrow band of electron energies of width about kT each side of the Fermi energy (we return to this point below).

Transport Properties of Solids

Then

Total average kinetic energy of participating conduction electrons in the absence of an applied electric field = Average thermal kinetic energy + Kinetic energy by virtue of the fact that they are F.D. electrons in a metal

Total average kinetic energy of conduction electrons in the absence of an applied field

$$\sim \bar{\varepsilon} + \varepsilon_F$$

where from equations (4.7b) and (6.5) we have

$$\bar{\varepsilon} = \tfrac{3}{2}kT = \tfrac{1}{2}m_e^*\overline{v^2}$$

$$\varepsilon_F = \frac{h^2}{8m_e^*}\left(\frac{3n_e}{\pi}\right)^{\frac{2}{3}}$$

Consider a temperature of 300°K. Then

$$\bar{\varepsilon} = \tfrac{3}{2}kT = 6\cdot21 \times 10^{-21} \text{ J} \tag{6.18}$$

and we have seen in Section 4.3 that for silver the Fermi energy is $9\cdot0 \times 10^{-19}$ J. Hence

$$\varepsilon_F \gg \bar{\varepsilon} \tag{6.19}$$

We can therefore write, with a good degree of approximation,

Total average kinetic energy of a conduction electron in the absence of an applied field

$$\sim \varepsilon_F$$

Let us introduce the Fermi speed v_{Fermi} defined by

$$\tfrac{1}{2}m_e^* v_{\text{Fermi}}^2 = \varepsilon_F \tag{6.20}$$

We note that the Fermi speed is *not* some average thermal speed; in fact it is not a *thermal* speed at all because it has a finite value even at the absolute zero. The Fermi speed is the participating conduction electron speed appropriate to the fact that the conduction electrons all have essentially the same energy, the Fermi energy. A slightly different way of phrasing this conclusion is to say that each and every conduction electron participating in any transport process must have an energy close to the Fermi energy and a speed close to the Fermi speed; we can think of the Fermi energy therefore as a 'zero point' energy for all conduction electrons. Over and above this zero-point energy, all conduction electrons participating in a transport process will also have an additional energy, a thermal energy $\bar{\varepsilon} = \frac{3}{2}kT$ and a root mean square thermal speed v_{th} where both the thermal energy and root mean square thermal speed are *very small* compared with the Fermi energy and the Fermi speed respectively. We therefore can conclude, perhaps rather surprisingly, that a heat transfer to a metal has an essentially negligible effect on the random motion of the conduction electrons.

Let us calculate the Fermi speed for silver. As we have seen in Section 4.3, there are 5.86×10^{28} valence electrons per cubic metre in silver and the Fermi energy is 9.0×10^{-19} J. If we take the effective mass of a conduction electron in silver equal to the isolated electron mass in free space (since we do not attempt to calculate the effective mass in this book) then from equations (4.7b) and (6.20) we have

$$v_{Fermi} = \frac{h}{2m_e^*} \cdot \left(\frac{3n_e}{\pi}\right)^{\frac{1}{3}} \tag{6.21}$$

$$= \frac{6.63 \times 10^{-34}}{2 \times 9.11 \times 10^{-31}} \left(\frac{3 \times 5.86 \times 10^{28}}{\pi}\right)^{\frac{1}{3}} \text{ ms}^{-1}$$

$$= 1.40 \times 10^6 \text{ m/s} \tag{6.22}$$

Experiment shows that the electrical conductivity of silver at $0°C$ is 6.80×10^7 $(\Omega m)^{-1}$. Hence, from equation (6.17)

$$\tau_{Fermi} = \frac{m_e^* \cdot \sigma}{n_e \cdot e^2}$$

$$= \frac{9.11 \times 10^{-31} \times 6.80 \times 10^7}{5.86 \times 10^{28} \times (1.60 \times 10^{-19})^2} \text{ s}$$

$$= 4.13 \times 10^{-14} \text{ s} \tag{6.23}$$

i.e. each conduction electron travels for only about 10^{-14} s between scattering collisions! We can also calculate the distance travelled by a conduction electron in this time, (the conduction electron mean free path)

$$l_{Fermi} = v_{Fermi} \cdot \tau_{Fermi}$$

$$= 1.40 \times 10^6 \times 4.13 \times 10^{-14} \text{ m}$$

$$= 5.75 \times 10^{-8} \text{ m} \tag{6.24}$$

which is about 70 lattice spacings, where a lattice spacing is the mean distance between any two nuclei of the lattice. This important result is partial proof of our earlier statement that conduction electrons would travel through a perfect lattice without meeting any 'resistance to motion'. We have now taken a real-life situation with silver, and shown that although the lattice is imperfect,

Transport Properties of Solids

nevertheless the conduction electrons can still travel no less than about 70 lattice spacings before suffering scattering collisions. Obviously, in the case of silver, although the lattice is imperfect, it is obviously not 'grossly imperfect' otherwise the conduction electrons would travel only a few lattice spacings before suffering scattering.

We must be careful in our use of equation (6.24)

$$l_{\text{Fermi}} = v_{\text{Fermi}} \cdot \tau_{\text{Fermi}}$$

Before F.D. statistics were discovered, it was thought that v, l, and τ were all temperature-dependent. However, our discussion of equation (6.19) has shown that v is in fact v_{Fermi} and is essentially constant, independent of temperature, so that l and τ are directly proportional and both of course refer to the 'Fermi energy' conduction electrons, that is, they are l_{Fermi} and τ_{Fermi}. The result that l_{Fermi} is about 70 lattice spacings is also further justification for our use of periodic boundary conditions for the wave equation that we discussed in Section 4.2 and Section 6.2 above. Even for the smallest laboratory sized specimens, say pieces 10^{-5} m in length, we see that this still represents about 10^3 lattice spacings. Hence, so far as a conduction electron is concerned the boundaries of the metal are effectively infinitely far away because every electron that reaches a metal boundary has suffered a very large number of collisions within the volume of the metal rather than having travelled freely from boundary to boundary as envisaged in Section 2.3.

We have now reached the stage where we cannot defer any longer a discussion of the lattice-imperfection scattering of the conduction electrons. Deviations from an exactly periodic lattice potential do occur. These deviations in the potential are due to:

a. Dynamic deviations where the nuclei are oscillating due to their thermal motion and are therefore not stationary in space. We have already discussed the thermal motion of the lattice in Chapter 5, both in terms of the discrete (Einstein) and continuum (Debye) models and shown their equivalence at 'high' temperatures.
b. Static deviations such as the presence of impurity and/or interstitial atoms or mechanical defects in the lattice such as dislocations.

In most metals at approximately room temperature the thermal motion of the lattice is by far the most important cause of deviation from a perfectly regular lattice. Hence the force that 'resists' the unimpeded motion of the conduction electrons in the applied field is due to electron–phonon scattering collisions, and this is why we discussed the phonon gas in Chapter 5.

Let us return to our discussion of Section 6.3 of the effect of the resistive force on the electrons and reconsider it in the light of the comment that electron–phonon collisions are responsible at and around room temperature. Such collisions are practically elastic; the electron does not lose much of its kinetic energy in colliding with, for example, a phonon. Analysis shows that when an electron is scattered by an elastic wave or phonon then the number of phonons in the elastic wave changes by ± 1. However, from Chapter 4 the energy of a phonon is $h\nu$. If we take the frequency ν equal to the Debye cut-off frequency ν_{max} then

$$\text{maximum energy of phonon} = h\nu_{max}$$
$$= k\theta_D \quad (6.25)$$

For silver, the Debye temperature is $226°K$ whence the maximum energy of a phonon in silver is $3 \cdot 1 \times 10^{-21}$ J. This is about the average thermal energy of a conduction electron [see equation (6.18)] but very much smaller than the 'zero-point' energy of a conduction electron with the Fermi energy of $9 \cdot 0 \times 10^{-19}$ J.

Consequently in an electron–phonon scattering collision, the electron loses only a very small amount of kinetic energy but changes its momentum from some initial value to a final value corresponding to this very slightly lower energy. Furthermore, although the conduction-electron energy (which, as we have shown, is essentially the Fermi energy) is much greater than even the maximum phonon energy, nevertheless both conduction electrons and phonons have similar values of momentum. Consequently in a conduction electron–phonon scattering collision although there is very little energy exchange (the collision being essentially elastic), there is a considerable momentum exchange. Hence, phonons are very effective in destroying electron momentum—and it is exactly electron momentum that constitutes electrical current flow in the direction of the applied electric field. Therefore phonons are a very important source of 'resistance to motion' of the conduction electrons.

Strictly speaking, this conclusion only holds for high-frequency phonons. At low electrical conductor temperatures where all of the phonons have a predominantly low frequency, the conduction electron scattering process is different. We shall consider this in our discussion of the Wiedemann–Franz–Lorenz law in Section 6.9. Returning to our consideration of the room-temperature electrical conductor, in a conduction electron–phonon scattering collision, the electron must be able to move into an empty translational energy state whose energy value is only very slightly less than the energy that the conduction electron had before collision. As we have seen in Section 6.3, most valence electrons therefore cannot undergo scattering because the Pauli principle disallows them to do so as there are no empty translational energy states available for them to move into. It is only those electrons close to the Fermi energy (the conduction electrons) that can find empty energy states available to 'drop into'. Hence, only the conduction electrons suffer scattering. We can summarize this situation by saying that *all* valence electrons are accelerated in the applied

Transport Properties of Solids

electric field because there is always an empty energy state ready to receive any valence electron which is changing its energy state (due to gaining energy from the applied field), the vacancy being created by the simultaneous change of energy state of another valence electron—and that this movement of *all* of the valence electrons takes place between time $t = 0$ when the electric field is first applied and the time τ. Thereafter, the 'resistive force' is felt by the electrons, thus electron–phonon scattering collisions commence but only for the conduction electrons near the Fermi level. On scattering these conduction electrons lose their drift velocity and the whole acceleration–resistance process starts all over again. Furthermore, in the scattering process the energy loss by the drifting conduction electrons is small because the electron only loses an energy equal to one phonon energy which we have just shown to be small compared with the Fermi energy.

We can use these results to demonstrate the fact that the drift velocity is considerably less than the thermal speed which in turn is considerably less than the Fermi speed. We can readily do this if we consider the drift energy of a conduction electron in the external electric field and compare it with the energy of a conduction electron moving at the Fermi level. Suppose the electric field is 10^4 V/m. In one free path the conduction electron will gain a potential energy equal to its charge multiplied by the potential through which it moves. In this case this is:

$$1\cdot 6 \times 10^{-19} \times 10^4 \times 5\cdot 8 \times 10^{-8} \text{ J} = 9\cdot 28 \times 10^{-23} \text{ J}$$

This energy is ultimately transferred to the lattice at a collision in the form of joule heat. Clearly, it is a very small fraction of both the thermal and Fermi energy of an electron at the Fermi level which in the case of silver are $6\cdot 21 \times 10^{-21}$ J and $9\cdot 0 \times 10^{-19}$ J respectively. We therefore picture the conduction electrons moving individually and randomly through the conductor with a high speed (essentially the Fermi speed) and yet all drifting slowly through the conductor in the direction of the applied electric field with the drift velocity such that

$$v_e \ll \overline{v_{\text{th}}} \ll v_{\text{Fermi}}$$

and

$$\text{Drift energy} \ll \bar{\varepsilon} \ll \varepsilon_{\text{Fermi}}$$

6.4 The effect of electron–phonon scattering collisions on the electrical conductivity of the valence-electron gas

In Chapter 5 we developed a simple thermal theory of the lattice in terms of the thermal oscillations of the solid (or in terms of the phonon gas). Let us see if we can use this theory to give us some idea of the effects of electron–phonon scattering collisions on the electrical conductivity. In this book we have made some approximate calculations and the reader should be warned that the following calculation is even more approximate than most!

The departure of a real lattice from a perfectly spaced array of lattice nuclei is primarily due to the fact that each volume element of the solid is not fixed in space but is oscillating about some equilibrium position. Each volume element is vibrating in three dimensions and such a motion can be regarded as equivalent to three one-dimensional quantum oscillators, each oscillating at right angles to the other two. Let us consider any one of these one-dimensional quantum oscillators. It has energy ε given by equation (5.4)

$$\varepsilon = \tfrac{1}{2}\mu v^2 + \tfrac{1}{2}f(x - x_e)^2$$

Since we are considering the *thermal* kinetic energy of the oscillator this becomes

$$\varepsilon = \tfrac{1}{2}\mu \overline{v^2} + \tfrac{1}{2}f \cdot (x - x_e)^2$$

where the first term on the right-hand side is the thermal kinetic energy because $\overline{v^2}$ is the mean square thermal speed [see Equation (6.6)]. We shall regard the oscillating volume element as representing an effective area 'blocking' the free movement of a conduction electron and therefore we need to know something about the amplitude of vibration of the oscillator since this amplitude will decide how 'big' the oscillating volume element looks to the conduction electron. Equation (5.8) shows that

$$\text{effective amplitude of oscillator} = A$$

$$= \left[\frac{\varepsilon}{2\mu(\pi\nu)^2}\right]^{\tfrac{1}{2}}$$

But from equation (5.12) we have in general (not only for thermal energy),

$$\text{energy of oscillator} = mh\nu$$

Hence for the problem of a thermal lattice we can write

$$\text{average } thermal \text{ energy of an oscillator} = \bar{\varepsilon}$$

$$= \bar{m}h\nu \qquad (6.26)$$

where $\bar{m}$ = average number of phonons per oscillator

$$= \frac{1}{\exp\dfrac{h\nu}{(kT)} - 1} \qquad (6.27)$$

where we have used equation (5.35)

Transport Properties of Solids

Hence

$$\text{average thermal energy of an oscillator} = \bar{\varepsilon} = \frac{h\nu}{\exp\dfrac{h\nu}{(kT)} - 1} \quad (6.28)$$

If $T > \theta_D$ then from the reasoning that led to equation (5.27) we can write

$$\bar{\varepsilon} \sim \frac{h\nu}{\left(1 + \dfrac{h\nu}{kT}\right) - 1}$$

$$\sim kT \quad (6.29)$$

This result could have been written down directly from equation (5.26) by merely dividing both sides of that equation by $3N$, the number of one-dimensional oscillators in the lattice. Hence,

$$A = \left[\frac{kT}{2\mu(\pi\nu)^2}\right]^{\frac{1}{2}} \quad (6.30)$$

At temperatures $T > \theta_D$ we have stated (see Section 5.5) that all of the $3N$ one-dimensional oscillators will be participating and a considerable number of these will be oscillating at frequencies close to $\nu_{\max}$ since $dg(\nu) \propto \nu^2$. In addition these high-frequency oscillators carry the greatest energy and have the greatest amplitude of oscillation [see equation (5.8)].

$$\nu \sim \nu_{\max} \sim \frac{k\theta_D}{h}$$

$$\therefore \quad A = \sqrt{\frac{kT}{2\mu}} \cdot \frac{1}{\pi} \cdot \frac{h}{k\theta_D}$$

The effective area Q_{eff} presented by an oscillating volume element to an approaching conduction electron is

$$Q_{\text{eff}} = \pi A^2 \quad (6.31)$$

$$= \frac{1}{\pi} \cdot \frac{h^2 T}{k\theta_D^2 \mu}$$

We note that $Q_{\text{eff}} = 0$ when $x = x_e$, thus there will be no scattering of the conduction electrons when the lattice-volume elements are in their equilibrium positions and this of course is the perfectly periodic lattice which gives infinite electrical conductivity as discussed in Section 6.2.

Kinetic theory tells us that the mean free path of a conduction electron moving through the lattice is

$$l \sim \frac{1}{nQ_{\text{eff}}} \tag{6.32}$$

where n is the number of scattering centres per unit volume. Since we are working in the temperature range above the Debye temperature, the Debye and the Einstein theories of the thermal motion of the lattice are the same (see Chapter 5) and consequently we can regard the oscillating volume elements of the lattice as the oscillating nuclei of the lattice (which of course is the original Einstein assumption). In this case n is the number of oscillating lattice nuclei per unit volume so that we can write that the mean free path is

$$l \sim \frac{\pi k \mu \theta_{\text{D}}^2}{nh^2 T} \tag{6.32a}$$

But from equations (6.17) and (6.24) we have, for the electrical conductivity,

$$\sigma = \frac{n_e \cdot e^2}{m_e^*} \cdot \frac{l_{\text{Fermi}}}{v_{\text{Fermi}}}$$

$$= \frac{n_e \cdot e^2}{m_e^*} \cdot \frac{\pi k \mu \theta_{\text{D}}^2}{nh^2 T v_{\text{Fermi}}}$$

$$= \frac{n_e}{n} \cdot \frac{e^2}{h^2} \cdot \frac{\mu}{m_e^*} \cdot \pi k \cdot \frac{\theta_{\text{D}}^2}{v_{\text{Fermi}}} \cdot \frac{1}{T} \tag{6.33}$$

we remember that n_e is the total concentration of valence electrons (see Section 6.3).

Let us use this formula to calculate the electrical conductivity of silver at 300°K. For the monovalent noble metals we have $n = n_e$ and for silver the Debye temperature is 226°K (see Table 5.2). The mass of a silver atom is the atomic weight of silver divided by the Avogadro number where the atomic weight of silver is 108 kg (kg mole)$^{-1}$. Also, the Fermi speed has been calculated in equation (6.22) to be 1.40×10^6 m/s. Hence

$$\sigma = \frac{(1.60 \times 10^{-19})^2}{(6.63 \times 10^{-34})^2} \cdot \frac{108\pi \times 1.38 \times 10^{-23}}{6.02 \times 10^{26} \times 9.11 \times 10^{-31}} \cdot \frac{226^2}{1.40 \times 10^6} \cdot \frac{1}{300} (\Omega \text{m})^{-1}$$

$$= 6.24 \times 10^7 (\Omega \text{m})^{-1}$$

The value at room temperature (see Table 6.1 below) is 6.21×10^7 $(\Omega \text{m})^{-1}$ which is astonishingly good agreement considering how over-simplified the calculation has been.

Of equal importance is the temperature-dependence of the electrical conductivity or better, the electrical resistivity of metals. We have the formula

Transport Properties of Solids

$$\frac{1}{\rho} = \sigma = \frac{n_e \cdot e^2}{m_e^*} \cdot \frac{l_{\text{Fermi}}}{v_{\text{Fermi}}} \tag{6.34}$$

The essential feature of metals (as opposed to semiconductors) is that n_e is a *constant independent of temperature*. In addition, as we have seen, v_{Fermi} is also essentially constant. Hence, the only factor in our expression for the electrical conductivity that can vary with temperature is l_{Fermi} which is the mean free path of the conduction electrons at and around the Fermi level.

In the thermodynamic analyses given in earlier chapters we were able to obtain accurate quantitative results for all the thermodynamic properties. The reason why this is so is quite a profound one in that the state of thermodynamic equilibrium is independent of the collisional processes that cause it and is describable (in principle) purely in terms of readily measured thermodynamic properties. However, in this chapter we are dealing with a state where thermodynamic equilibrium in the thermodynamic system does not exist because there is a gradient of some thermodynamic property like electric field or temperature within the system. Consequently, when we discuss such transport processes we must evaluate the various collisional (scattering) mechanisms that produce the steady-state condition referred to in Section 6.1. However, a full discussion of electron and phonon scattering mechanisms is a problem in quantum-scattering theory and a textbook on thermodynamics is no place to develop such a theory, which can be rather complicated if done properly. We therefore will have to be content with simple semiquantitative calculations at best and qualitative discussions where such simple calculations cannot be given at all. Nevertheless such a simplified approach can give the essential features of the physical behaviour of metals and is therefore useful. We shall see that the main problem that will concern us is a determination of the variation of the mean free path of the conduction electrons having the Fermi energy (l_{Fermi}) with temperature or any other thermodynamic property such as externally applied pressure or externally applied electric field.

Let us therefore return to our consideration of the temperature dependence of the electrical resistivity of a metal. Equation (6.33) shows that in the region $T > \theta_D$ we have

$$\rho \propto \frac{T}{\mu \theta_D^2}$$

i.e. the electrical resistivity of a metal increases linearly with increasing temperature. The reader will have met this before in a slightly different guise where the temperature variation of the electrical resistance R_r of a metal is linear over a wide range of temperature. We normally write this as

Temperature coefficient of electrical resistance

$$= \frac{R_r(T_2) - R_r(T_1)}{R_r(T_1) \cdot [T_2 - T_1]} \qquad (6.35)$$

Here, $R_r(T)$ is the electrical resistance of the metal evaluated at some temperature T. Table 6.1 gives some values of this temperature coefficient of resistance.

The reason why the electrical resistivity increases with increasing temperature can be seen from equation (5.38), viz.

$$\text{number of phonons in a solid} = \frac{9NT}{2\theta_D}$$

Consequently as the temperature of the conductor increases, so does the number of phonons (see Section 5.7) which of course are the conduction electron

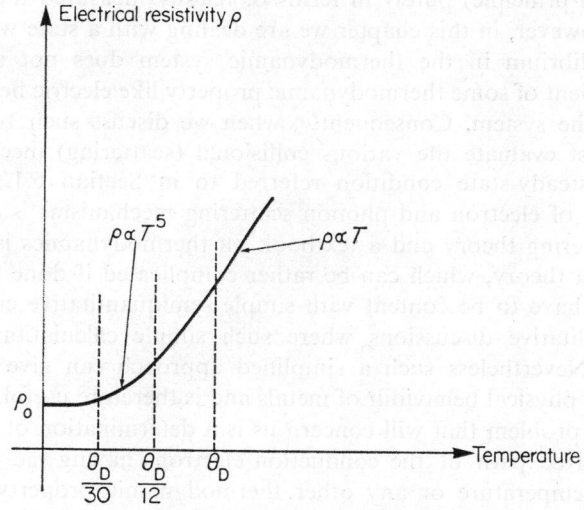

Figure 6.1. Schematic plot of the variation of electrical resistivity of a typical metal as a function of temperature (both scales are non-linear). ρ_0 is the residual resistivity.

scattering centres. Hence the drift motion of these conduction electrons in the externally applied electric field is impeded by this increasing number of phonons with the result that the resistivity increases.

Let us see what happens to the electrical resistivity as the temperature is lowered below the region described by equation (6.33) which is the region where the resistivity is linearly proportional to temperature. When the temperature falls to about $0.08\theta_D$ we find that the resistivity falls as T^5 rather than T. This result can be deduced from equation (5.39) together with a discussion of the effect on the mean free path of small-angle scattering collisions between

Transport Properties of Solids

electrons and phonons. We will not pursue this, however, but merely note that even in this T^5 region the electrical resistivity is still dominated by electron–phonon collisions. In practice the region of transition between a T and T^5 dependence of electrical resistivity may not be easily detected if the metal is impure. The reason is that at temperatures below about $0.03\theta_D$ the electrical resistivity ceases to vary with temperature at all and becomes constant. This occurs because at these low temperatures the resistivity ceases to be determined by electron–phonon scattering collisions at all because the phonon density has fallen to too low a value. When phonons cease to be the important scattering centres, then the scattering mechanism that limits the value of l_{Fermi} is the scattering of conduction electrons by impurity atoms and lattice defects.

Figure 6.1 shows a schematic variation of electrical resistivity with temperature for a typical metal, ignoring the phenomenon of super conductivity. The constant value of resistivity at low temperature is known as the residual resistivity ρ_0; clearly it depends upon the concentration of impurities or lattice defects but not the temperature (provided of course that the lattice defects do not change with temperature). For pure annealed metals ρ_0 is a small fraction of the total room-temperature resistivity. We can readily give a rough calculation of the residual resistivity because we have

$$\rho_0 = \frac{m_e^* \cdot v_{\text{Fermi}}}{n_e \cdot e^2 \cdot l_{\text{Fermi}}} \quad (6.36)$$

Here l_{Fermi} is the mean free path for electron-impurity atom-scattering collisions, i.e.

$$l_{\text{Fermi}} = \frac{1}{n_{\text{imp}} \cdot Q_{\text{imp}}}$$

Here we have used equation (6.32); n_{imp} is the concentration of impurities and Q_{imp} is the collision cross-section of an impurity atom when 'seen' by a conduction electron. Let us take the diameter of such an impurity atom to be 3×10^{-10} m

$$\rho_0 = \frac{m_e^* \cdot v_{\text{Fermi}} \cdot Q_{\text{imp}}}{e^2} \cdot \frac{n_{\text{imp}}}{n_e} \quad (6.37)$$

Consider the case of silver, discussed earlier,

$$m_e^* = m_e = 9 \cdot 11 \times 10^{-31} \text{ kg}$$
$$v_{\text{Fermi}} = 1 \cdot 40 \times 10^6 \text{ ms}^{-1}$$
$$e = 1 \cdot 60 \times 10^{-19} \text{ coulomb}$$
$$Q_{\text{imp}} = \tfrac{1}{4}\pi(3 \times 10^{-10})^2 \text{ m}^2$$
$$\frac{n_{\text{imp}}}{n_e} = 10^{-5},$$

i.e. there is one part in 10^5 atomic impurity.

Then the residual resistivity is

$$\rho_0 = \frac{9 \cdot 11 \times 10^{-31} \times 1 \cdot 40 \times 10^6 \times 0 \cdot 25\pi \times 9 \times 10^{-20}}{(1 \cdot 60 \times 10^{-19})^2} \cdot \frac{1}{10^5} \, \Omega\text{m}$$

$$= 3 \cdot 5 \times 10^{-11} \Omega\text{m}$$

which is a reasonable value for spectrographically standardized silver (see Table 6.1).

This procedure can be inverted and in fact represents an important practical application of the theory of the electrical resistivity of metals; since the residual resistivity is sensitive to the degree of purity of a metal we can obtain information on this degree of purity by simply measuring its electrical resistivity! For chemically very pure metals of known impurity content (determinable spectrographically or mass-spectrometrically) we can likewise study lattice-defect populations by studying the electrical resistivity.

Nowadays the technology of thin metallic films is also of great importance and we can discuss their behaviour qualitatively using the analysis already given. If we lower the temperature of such a thin film down towards the absolute zero and ensure that the metallic film is chemically pure, we find that the value of l_{Fermi} ceases to depend on the impurity or lattice defect content and becomes limited by the actual boundaries of the metal itself. We have already calculated [equation (6.24)] that for silver at room temperature $l_{\text{Fermi}} = 5 \cdot 8 \times 10^{-8}$ m. The resistivity of pure metals can decrease by a factor of several thousand at very low temperatures meaning that l_{Fermi} can become as large as 10^{-5} m or more—and we can readily construct thin metallic films of this 'thickness' or less. (Of course if we lower the temperature towards a few degrees above absolute zero, electrical superconductivity may arise but we shall not consider this at all in this book.)

Hence, ignoring superconductivity, we can write the electrical resistivity of a metal as

$$\rho = \rho_0 + \rho_i(T) \tag{6.38}$$

This 'adding-up' of resistivities is Matthiesen's rule; here ρ_0 is the temperature-independent residual resistivity due to impurities and lattice defects, that is, due solely to physical and chemical imperfections. $\rho_i(T)$ is the temperature-dependent resistivity known alternatively as either the 'intrinsic', 'ideal', 'thermal', 'lattice' or 'phonon' resistivity. It is due entirely to electron–phonon scattering collisions and is characteristic of an ideally chemically pure metal with a physically perfect lattice. Table 6.1 gives some values of ρ_0 and ρ_i where the latter is given as a function of temperature. In all cases the units are Ωm and residual resistivities have been subtracted from the $\rho_i(T)$ values.

Table 6.1 The intrinsic electrical resistivity and residual resistivity of common conductors in Ωm tabulated as a function of temperature: each entry has been multiplied by a factor of 10^8. The final column gives the temperature coefficient of resistance at 0°C.

Temperature °K	20	50	100	150	200	273	295	Typical value of residual resistivity	Temperature coefficient of resistance at 273°K in $(\deg C)^{-1} \times 10^3$
Alkali metals									
Na	1.7×10^{-2}	3.2×10^{-1}	1.2	2.0	2.9	4.3	4.8	1.0×10^{-3}	5.5
K	1.1×10^{-1}	7.2×10^{-1}	1.8	3.0	4.3	6.5	7.2	5.0×10^{-3}	5.4
Cs	8.8×10^{-1}	2.7	5.6	8.8	12.2	18.0	20.0	4.0×10^{-2}	5.0
Noble metals									
Cu	8.0×10^{-4}	5.0×10^{-2}	3.5×10^{-1}	7.1×10^{-1}	1.1	1.6	1.7	3.0×10^{-2}	4.3
Ag	3.8×10^{-3}	1.1×10^{-1}	4.2×10^{-1}	7.4×10^{-1}	1.0	1.5	1.6	1.0×10^{-3}	4.1
Au	1.3×10^{-2}	2.0×10^{-1}	6.3×10^{-1}	1.0	1.4	2.0	2.2	1.0×10^{-3}	4.0
Miscellaneous metals									
Zn	5.2×10^{-2}	4.9×10^{-1}	1.6	2.7	3.8	5.5	5.9	5.0×10^{-3}	4.3
Cd	1.3×10^{-1}	8.7×10^{-1}	2.3	3.6	4.9	6.7	7.3	1.0×10^{-4}	4.1
Al	6.0×10^{-4}	5.0×10^{-2}	4.7×10^{-1}	1.1	1.7	2.5	2.7	1.0×10^{-4}	4.7
Pb	5.6×10^{-1}	2.8	6.5	10.2	13.9	19.3	21.0	4.0×10^{-4}	4.2
Bi	5.8	19.0	3.7	55.0	74.0	105.0	116.0	4.0×10^{-1}	
Mo		1.1×10^{-1}	1.0	2.1	3.2	4.8	5.3	7.0×10^{-2}	5.3
W	5.6×10^{-3}	1.5×10^{-1}	9.2×10^{-1}	2.1	3.2	4.8	5.3	3.0×10^{-3}	4.8
Fe	7.0×10^{-3}	1.4×10^{-1}	1.2	3.1	5.3	8.7	9.8	1.0×10^{-2}	6.8
Ni	9.0×10^{-3}	1.5×10^{-1}	1.0	2.2	3.7	6.2	7.0	2.0×10^{-2}	6.8
Pt	3.6×10^{-2}	7.2×10^{-1}	2.7	4.8	6.8	9.6	10.4	4.0×10^{-3}	3.9

So far the discussion has been restricted to the room temperature and low-temperature behaviour of the electrical resistivity. We have already noted from equation (6.33) that for temperatures greater than about θ_D the electrical resistivity is directly proportional to the temperature. This linear behaviour given by equation (6.33) ceases at high temperatures, say above 1,000°K. This follows in fact from equation (6.33), viz.

$$\rho \propto \mu^{-1} . T . \theta_D^{-2}$$

where we recall that $\theta_D = \dfrac{h\nu_{max}}{k}$. At high temperatures ν_{max} becomes a function of temperature due to thermal expansion of the lattice. In Section 5.5 we noted that the Debye harmonic oscillator model did not allow for thermal expansion and in Section 6.7 we shall discuss the anharmonic oscillations of the lattice. In that section we shall find that the quartic term in the potential energy expression 'softens' the anharmonic oscillations at large amplitudes by reducing the energy necessary to produce a displacement from equilibrium. Hence thermal expansion reduces the effective storing force in the oscillation, reduces the frequency of oscillation of the nuclei and therefore in turn reduces the Debye temperature. This follows from equation (5.19), viz.

$$\theta_D = \frac{h\nu_{max}}{k} = \frac{h}{k}\left[\frac{9N}{4\pi V} \cdot \frac{1}{\dfrac{1}{a_L^3}+\dfrac{2}{a_T^3}}\right]^{\frac{1}{3}}$$

i.e.
$$\theta_D \propto V^{-\frac{1}{3}} \qquad \text{6.39a}$$

Hence
$$\rho \propto \mu^{-1} . T . V^{\frac{2}{3}} \qquad \text{(6.39b)}$$

Therefore the electrical resistivity increases more rapidly than a linear dependence on the temperature due to thermal expansion of the conductor (increase of volume V).

A more complete analysis than we can give in this book would involve a detailed study of the effect of both the temperature and pressure dependence of l_{Fermi}. For example, for most metals the electrical conductivity increases with increasing external pressure on the metal (although there are some exceptions such as lithium and bismuth); we may qualitatively expect this from equation (6.33) since

$$\sigma \propto \mu . \theta_D^2 . T^{-1}$$

∴
$$\frac{d\sigma}{dp} \propto \frac{d(\theta_D^2)}{dp} > 0$$

Transport Properties of Solids

Physically this arises because at high external pressures the volume of the metal decreases. This means the force between neighbouring nuclei increases and therefore the Debye temperature decreases following equation (6.39a). Alternatively equation (6.39b) shows that the electrical resistivity decreases (and therefore the electrical conductivity increases) as the volume decreases due to the external pressure. Strain gauges operate using the principle that the resistivity of a metal is dependent on its state of stress (tension rather than compression).

Let us now turn from considering the electrical conductivity of metals to a consideration of their thermal conductivity as defined in equation (6.3). The problem here is somewhat easier because, as we shall demonstrate, we can find without difficulty a formula that gives the thermal conductivity in terms of a thermodynamic property that we have already determined in earlier chapters.

6.5 A relationship between thermal conductivity and heat capacity

In this section we shall link the thermal conductivity and constant-volume heat capacity of any thermally perfect gas. Clearly this is an important relationship since we have already calculated the constant-volume heat capacity of a plasma, a valence-electron gas and a phonon gas in earlier chapters.

Consider a thermally perfect gas confined to a container of volume V. Let us draw three parallel planes in the gas (see Figure 6.2) such that the top plane is at a temperature T_1 and is separated by one mean free path l from the middle

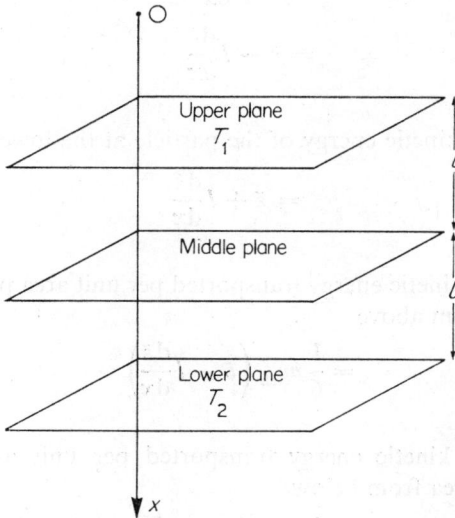

Figure 6.2. The three planes used in deriving the formula linking thermal conductivity with heat capacity.

plane; likewise the third plane is at temperature T_2 and separated by one mean free path from the middle plane. The transport process of thermal energy or, more colloquially, 'heat conduction' is due to the fact that although particles cross the middle plane from both above and below in *equal* numbers, those from above carry more thermal energy than those from below because the former come from regions of higher energy (represented by the higher temperature) than the latter; this is so because the downward-moving particles coming from the region bounded by the upper plane at T_1 and the middle plane will, on average, not suffer a collision until they have passed through the middle plane. Likewise we can draw a similar conclusion for those upward-moving particles from the region bounded by the lower plane and the middle plane.

On average one-sixth of the particles will be moving downward through the middle plane and one-sixth moving upward. If the root mean square thermal speed of the particles is $\overline{v_{\text{th}}}$ and if there are n particles per unit volume then

Number of particles crossing unit area of the middle plane per unit time in the downward direction

$$= \tfrac{1}{6} n \, \overline{v_{\text{th}}} \qquad (6.40)$$

Let $\bar{\varepsilon}$ be the average thermal kinetic energy of any particular particle at the middle plane. Then

Average thermal kinetic energy of the particle at the upper plane

$$= \bar{\varepsilon} - \frac{d\bar{\varepsilon}}{dx}.\, dx_l$$

$$= \bar{\varepsilon} - l\frac{d\bar{\varepsilon}}{dx}$$

Similarly,

Average thermal kinetic energy of the particle at the lower plane

$$= \bar{\varepsilon} + l\frac{d\bar{\varepsilon}}{dx}$$

Hence,

Average thermal kinetic energy transported per unit area per unit time across the middle plane from above

$$= \frac{1}{6} n\overline{v_{\text{th}}} \left(\bar{\varepsilon} - l\frac{d\bar{\varepsilon}}{dx} \right)$$

and

Average thermal kinetic energy transported per unit area per unit time across the middle area from below

$$= \frac{1}{6} n\overline{v_{\text{th}}} \left(\bar{\varepsilon} + l\frac{d\bar{\varepsilon}}{dx} \right)$$

Transport Properties of Solids

Consequently,

Net flux of thermal kinetic energy downwards through the middle plane

$$= \frac{1}{6} n \overline{v_{th}} \left[\left(\bar{\varepsilon} - l \frac{d\bar{\varepsilon}}{dx} \right) - \left(\bar{\varepsilon} + l \frac{d\bar{\varepsilon}}{dx} \right) \right]$$

$$= -\frac{1}{3} n \overline{v_{th}} l \frac{d\bar{\varepsilon}}{dx} \qquad (6.41)$$

But from equation (6.3):

$$\text{Flux of thermal kinetic energy} = -\lambda \frac{dT}{dx}$$

∴

$$-\lambda \frac{dT}{dx} = -\frac{1}{3} n \frac{d\bar{\varepsilon}}{dx} \overline{v_{th}} l$$

i.e.

$$\lambda \frac{dT}{dx} = \frac{1}{3} n \frac{d\bar{\varepsilon}}{dT} \frac{dT}{dx} \overline{v_{th}} l$$

i.e.

$$\lambda = \frac{1}{3} \frac{d(n\bar{\varepsilon})}{dT} \overline{v_{th}} l$$

But the total thermal kinetic energy of all of the particles is

$$E = nV\bar{\varepsilon}$$

since

$$\bar{\varepsilon} = \frac{E}{N} = \frac{E}{nV}$$

∴

$$\frac{d(n\bar{\varepsilon})}{dT} = \frac{1}{V} \frac{dE}{dT}$$

$$= \frac{C_V}{V}$$

∴

$$\lambda = \frac{1}{3} \cdot \frac{C_V}{V} \cdot \overline{v_{th}} l \qquad (6.42)$$

which relates the thermal conductivity to the constant-volume heat capacity.

We shall find that our discussions in the following sections concerned with the application of equation (6.42) to electrons in solids are at best only semi-quantitative. We have already given a reason for this in Section 6.4; namely a determination of the electron or phonon mean free path l, is a difficult problem in quantum mechanics and not appropriate for discussion in a textbook on thermodynamics. Although we could immediately apply equation (6.42) to determine the thermal conductivity of the plasma discussed in Chapter 3, we shall not do so since the reader can readily perform this calculation which raises no new problems. Rather, it is more interesting to apply equation (6.42) to solids using the results from Chapters 4 and 5 that describe the valence-electron gas and phonon gas.

We have already shown in this chapter that in the case of the valence-electron gas in a metal at and around room temperature the conduction electrons moving under the influence of an externally applied electric field are 'resisted' in their movement by suffering scattering collisions with phonons. In the present case of an externally applied temperature difference (which of course is the cause of thermal conduction) we shall find that once again conduction electrons in a metal are impeded by scattering collisions with phonons. We might therefore expect the Debye temperature once again to represent a characteristic temperature for our discussion. What we shall in fact do is to discuss the asymptotic values of thermal conductivity (both valence electron and lattice) for 'high' temperatures, i.e. $T > \theta_D$ and 'low' temperatures, i.e. $T \ll \theta_D$ using simple approximations for the mean free path l. The intermediate temperature range will not be quantitatively analysed at all—just as it was neglected in Section 5.5 where our discussion of Debye theory of the lattice was restricted to its limiting forms of high and low temperatures.

6.6 The thermal conductivity of the valence-electron gas

We have already remarked in Section 6.3 that the valence-electron gas considered in Chapter 4 cannot explain transport processes at all, because such transport processes are due to collisional interactions between the conduction electrons and the lattice, and the valence-electron gas model completely ignores the lattice!

If we allow for the existence of electron–phonon collisions then we can readily find the thermal conductivity of the valence-electron gas. Equation (6.42) is, for valence electrons,

$$\lambda_e = \frac{1}{3} \cdot \frac{C_V^{\text{electrons}}}{V} \cdot (\overline{v_{\text{th}}})_e \cdot l_e$$

We must be careful how we interpret $(\overline{v_{\text{th}}})_e$ in this formula. In our analysis in Section 6.5, $\overline{v_{\text{th}}}$ arose as the root mean square *thermal* speed of the particles

Transport Properties of Solids

since all particles were assumed to have no other speed than their thermal speed due to their random motion. However, in the present case of the valence electrons in a metal we know from our analysis in Section 6.3 that only those valence electrons close to the Fermi level (the conduction electrons) actually participate in the thermal conduction process (since they alone can find empty translational quantum states to 'jump into'). Also, we showed in that section [see equation (6.19)] that the total kinetic energy of a conduction electron was effectively ε_F since its average *thermal* kinetic energy was very much smaller than its Fermi energy so that the former could be ignored. Also, from equation (6.20) the Fermi speed was

$$v_\text{Fermi} = \sqrt{\frac{2\varepsilon_\text{F}}{m_e^*}}$$

Hence
$$\overline{(v_\text{th})_e} \sim v_\text{Fermi}$$

In addition,
$$l_e \equiv l_\text{Fermi}$$

$\therefore$

$$\lambda_e = \frac{1}{3} \cdot \frac{C_V^\text{electrons}}{V} \cdot v_\text{Fermi} \cdot l_\text{Fermi} \quad (6.43)$$

where
$$l_\text{Fermi} = v_\text{Fermi} \cdot \tau_\text{Fermi}$$

Now from equation (4.19)

$$C_V^\text{electrons} = \frac{\pi^2}{2} k N_e \left(\frac{kT}{\varepsilon_\text{F}}\right)$$

$\therefore$

$$\lambda_e = \frac{\pi^2}{6} \cdot \frac{N_e}{V} \frac{k^2 T}{\varepsilon_\text{F}} \cdot v_\text{Fermi} \cdot l_\text{Fermi} \quad (6.44)$$

or, using equation (6.20) and putting $n_e = \frac{N_e}{V}$

$$\lambda_e = \frac{\pi^2}{3} \frac{k^2 T}{m_e^*} \cdot n_e \cdot \tau_\text{Fermi} \quad (6.45)$$

Once again we see that the parameter under discussion, λ_e depends upon the total concentration of valence electrons n_e and not just on the concentration of conduction electrons.

Let us examine the asymptotic behaviour of this formula for the valence-electron gas thermal conductivity in the limits of 'high' and 'low' temperature, $T > \theta_D$ and $T \ll \theta_D$ respectively assuming that phonons are the scattering centres for the conduction electrons.

High-temperature limit. $T > \theta_D$

Equation (6.44) is

$$\lambda_e = \frac{\pi^2}{6} n_e \frac{k^2 T}{\varepsilon_F} v_{\text{Fermi}} \cdot l_{\text{Fermi}}$$

i.e.

$$\lambda_e \propto \frac{n_e \cdot T}{\varepsilon_F} v_{\text{Fermi}} \cdot \frac{1}{n_{\text{scat}}} \tag{6.46}$$

where we have used equation (6.32) for the mean free path. Here n_{scat} is the number of scattering centres per unit volume and these scattering centres are phonons. Equation (5.38) showed that in the high-temperature limit of $T > \theta_D$

$$n_{\text{scat}} = \frac{N_p}{V}$$

$$= \frac{9NT}{2\theta_D \cdot V} = \frac{9nT}{2\theta_D}$$

$$\therefore \quad \lambda_e \propto \frac{n_e}{n} \cdot \frac{v_{\text{Fermi}} \cdot \theta_D}{\varepsilon_F} \tag{6.47}$$

To a first approximation, all of the quantities on the right-hand side of the proportionality in equation (6.47) are independent of temperature Hence if $T > \theta_D$ then,

$$\lambda_e = \text{constant} \tag{6.48}$$

Let us denote this limiting value of the phonon component of the valence-electron gas thermal conductivity at high temperature by the value $(\lambda_e)_\infty$.

Low-temperature limit $T \ll \theta_D$

Equation (5.39) showed that for $T \ll \theta_D$ we have that the number of phonons per unit volume is

$$n_{\text{scat}} = 21 \cdot 64 \frac{N}{V} \left(\frac{T}{\theta_D}\right)^3$$

Hence from equation (6.46)

$$\lambda_e \propto \frac{n_e}{n} \frac{v_{\text{Fermi}}}{\varepsilon_F} \frac{\theta_D^3}{T^2} \cdot \frac{1}{T^2} \tag{6.49}$$

Transport Properties of Solids

If we assume, to a first approximation, that n_e, n, v_{Fermi}, θ_D and ε_F are all sensibly independent of temperature then

$$\lambda_e \propto \frac{1}{T^2} \qquad (6.50)$$

Experiment shows that this result holds quite accurately in the temperature region

$$\frac{\theta_D}{20} < T < \frac{\theta_D}{5}$$

This means that the phonon component of the electron gas thermal conductivity *increases* as the temperature decreases. This increase will continue until the phonon density is so low that the mean free path of the conduction electrons is no longer dictated by conduction electron–phonon scattering collisions. If the metal is chemically pure and free of lattice defects then the conduction electrons are scattered by the actual boundaries of the metal; they effectively travel the entire length of the metal without scattering but then scatter at the edges of the metal. Clearly in this very low temperature limit, the conduction electron mean free path is equal to the distance between the faces of the metal l_{solid}. (This is entirely analogous to the electrical conductivity case discussed in Section 6.4 when we discussed thin metallic films). When these conditions prevail we have

$$\lambda_e = \frac{1}{3} \cdot \frac{C_V}{V} \cdot v_{\text{Fermi}} \cdot l_{\text{solid}}$$

i.e.

$$\lambda_e \propto C_V \qquad (6.51)$$

i.e.

$$\lambda_e \propto T^3 \qquad (6.52)$$

where the latter result follows from Debye low-temperature theory [see equation (5.30)]. Hence we conclude that

$$\lambda_e \to 0 \quad \text{as} \quad T \to 0 \qquad (6.53)$$

Equation (6.52) is experimentally found to hold in the temperature region $T < \frac{\theta_D}{30}$.

What happens in the intermediate temperature region $T \leqslant \theta_D$? The answer is that equations (6.48) and (6.50) *also* cover this interval, such that as the temperature of the conductor is reduced from θ_D the phonon component of the valence-electron gas thermal conductivity ceases to be constant [as given by equation (6.48)] and increases [as given by equation (6.50)]. This increase

168   *Thermodynamics of Electrical Processes*

is to be expected since equation (6.46) shows that the thermal conductivity is inversely proportional to the phonon density. This density of scattering centres decreases with temperature decrease [see equation (5.38)] and consequently the conduction electron mean free path increases. The phonon component of the valence-electron gas thermal conductivity continues to increase and eventually reaches a maximum at a temperature of about $\frac{\theta_D}{20}$; thereafter it falls according to equation (6.52). Figure (6.3) shows a schematic picture of the temperature

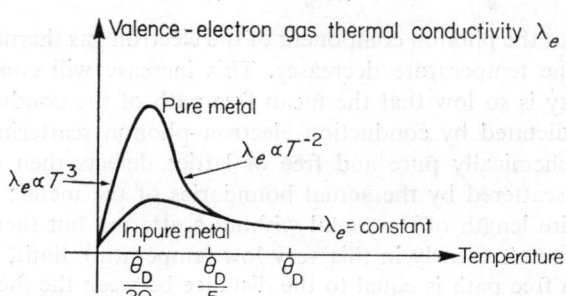

Figure 6.3. Schematic plot of the variation of valence-electron gas thermal conductivity λ_e of a pure and of an impure metal as a function of temperature (the temperature scale is non-linear).

variation of the phonon component of the valence-electron gas thermal conductivity of a metal; for a pure metal the maximum value is about ten times the room-temperature value.

In our discussion of electrical conductivity in Section 6.4 we noted the significant effects of impurities and lattice defects, especially at the lower temperatures. Similarly, in the present case, if the metal contains appreciable impurities and/or lattice defects, the behaviour of the valence-electron gas thermal conductivity is different as the temperature falls from the maximum at $\frac{\theta_D}{20}$. The low-temperature behaviour where $T < \frac{\theta_D}{20}$, say, differs from the behaviour given by equation (6.50) because at these lower temperatures the number of phonons per unit volume may be small compared with the number of impurity atoms or lattice defects per unit volume. In this case we have, analogously to the discussion supporting equation (6.36),

$$\lambda_e^{\text{imp}} \propto \frac{1}{3} \frac{1}{V} \left[\frac{\pi^2}{2} N_e k \left(\frac{kT}{\varepsilon_F} \right) \right] v_{\text{Fermi}} \frac{1}{n_{\text{imp}}}$$

i.e.

$$\lambda_e \propto \frac{n_e}{n_{\text{imp}}} \cdot \frac{T \cdot v_{\text{Fermi}}}{\varepsilon_F} \qquad (6.54)$$

i.e.

Hence, if $n_e, n_{\text{imp}}, v_{\text{Fermi}}$ and ε_F are independent of temperature then for $T < \dfrac{\theta_D}{20}$

$$\lambda_e \propto T \qquad (6.55)$$

Let us invert this expression and write it in terms of the impurity atom component of the valence electron gas thermal resistivity.

$$W_0 = \frac{1}{\lambda_e^{\text{imp}}} = \frac{\beta}{T} \qquad (6.56)$$

Here β is the impurity coefficient that depends upon the particular specimen being studied. β is very sensitive to the impurity level and the previous thermal history of the metal sample. The presence of even 1% of impurities alloyed with the metal can reduce the conduction electron mean free path by one order of magnitude or more below the value for a pure metal. In section 6.9 we shall show how to calculate values of β. The effect of such impurities results in a flattening of the maximum in the thermal-conductivity curve shown in Figure 6.3. For very impure metals the electron-impurity atom-scattering process dominates at all temperatures and the 'suppressed' lattice thermal conductivity which we shall discuss in Section 6.7 may become important.

If then we are dealing with impure metals the total valence-electron gas thermal resistivity for temperatures less than about $\dfrac{\theta_D}{20}$ is given by the sum of the electron–phonon and electron-impurity atom components.

$$W_e = \frac{1}{\lambda_e} = W_0 + W_i(T) = \frac{\beta}{T} + \alpha T^2 \qquad (6.57)$$

This equation is analogous to Matthiesen's rule [see equation (6.38)].

Here the term $W_i(T)$ equal to αT^2 is obtained from inverting equation (6.50). As we have noted above, W_0 is the temperature-independent impurity atom and/or lattice defect component of the valence-electron gas thermal resistivity, that is due solely to physical and chemical imperfections. $W_i(T)$ is the temperature-dependent phonon component of the valence-electron gas thermal resistivity known as the 'intrinsic' or 'ideal' or 'thermal' or 'phonon' valence-electron gas thermal resistivity since it is due entirely to electron–phonon scattering collisions and is characteristic of an ideally chemically pure conductor with a physically perfect lattice; α should be a constant for a particular metal.

If we examine equation (6.57) then provided β is small, there should be a minimum in W_e and therefore a maximum in λ_e. Clearly the purer the sample the smaller the value of β and the larger the maximum; this point is brought out in Figure 6.3. If we plot $W_e T$ versus T^3 then we should obtain a straight line (for those low temperatures when $T < \dfrac{\theta_D}{20}$) whose slope is α and whose intercept on the $W_e T$ axis gives β. Typical values of α, $(\lambda_e)_\infty$ are shown in Table 6.2.

Table 6.2 Values of the theoretical and experimental 'intrinsic,' that is phonon component of the valence-electron gas thermal resistivity coefficient α deduced from low-temperature studies. The table also gives values of the 'high' temperature ($T > \theta_D$) phonon component of the valence-electron gas thermal conductivity $(\lambda_e)_\infty$ of a selection of metals.

Metal	α_{exp} m (watt deg C)$^{-1}$	$\dfrac{\alpha_{\mathrm{exp}}}{\alpha_{\mathrm{theory}}}$	$(\lambda_e)_\infty$ watt (m deg C)$^{-1}$
Na	3.8×10^{-6}	0.21	135
K	1.2×10^{-5}	0.19	100
Cs	2.2×10^{-4}	0.32	33
Cu	2.5×10^{-7}	0.17	385
Ag	6.4×10^{-7}	0.18	420
Au	1.3×10^{-6}	0.17	310
Zn	3.0×10^{-6}		
Cd	4.0×10^{-6}		
Al	2.7×10^{-7}		230
Pb	2.5×10^{-5}		35
Bi			8
W	9.0×10^{-7}		
Fe	1.0×10^{-6}		63
Ni			90
Pt	4.3×10^{-6}		80

Of course, this discussion of the behaviour of the valence-electron thermal conductivity is only 'part of the story'. We must also allow for the effect of the 'rest' of the solid viz. the lattice thermal conductivity.

6.7 The thermal conductivity of the lattice phonon gas

In Chapter 5 we discussed the Debye theory of the heat capacity of a lattice. This theory is based on the idea that the thermal energy is stored in a solid

Transport Properties of Solids

as thermal elastic waves and that very small volume elements of the solid undergo simple harmonic motion (the reader will recall that the model was essentially an elastic continuum one where the 'atomicity' of the lattice was ignored other than in limiting the number of elastic waves). We can readily picture the process of thermal-energy transport in such a theory as one in which every small-volume element oscillates about its equilibrium position with an amplitude determined by the temperature [see equation (6.30)]. It interacts with its neighbouring small-volume elements and if these neighbouring volume elements were initially oscillating with a smaller amplitude characteristic of a lower temperature then energy will be transferred and the amplitude of oscillation of the neighbouring volume element will increase. Hence, if the ends of our solid are kept at different temperatures, thermal energy will be transferred in this manner from the hot to the cool face—and this is the process of thermal conduction. Unfortunately it is too good a picture of thermal conduction because we recall from Chapter 5 that these thermal elastic waves form a set of standing waves of $3N$ different frequencies and that such standing waves are independent of each other, that is, waves of different frequencies do not interact with each other. Consequently Debye theory allows for no 'resistance to motion' of these thermal elastic waves, so that thermal energy enters the solid at the hot face and travels without interruption to the cool face. If the solid is chemically pure and mechanically perfect and effectively infinite in extent, such that surface boundaries do not need to be considered, then the thermal conductivity of such a solid would be infinite since we would have direct thermal energy transfer. In many ways this is analogous to the movement of an electron through a perfect lattice when an electric field is applied to the conductor: in Section 6.2 we noted that this gave rise to infinite electrical conductivity. In the present case the thermal elastic wave is merely another type of wave motion that also propagates without scattering or attentuation directly through the volume of the solid.

Of course this cannot be the picture because we know from equation (6.3) (which is based on laboratory observation) that the flux of thermal energy through a solid is $-\lambda \dfrac{dT}{dx}$ and that the thermal conductivity λ is not infinite. In fact equation (6.3) indirectly tells us that the transport of thermal energy through a solid must be a diffusion-type process where the carriers of thermal energy, the thermal elastic waves or the phonons suffer scattering collisions as they descend the temperature gradient $\dfrac{dT}{dx}$. If such scattering collisions did not exist then the flux of thermal energy would depend upon $T_{\text{hot}} - T_{\text{cold}}$ and not upon the *gradient* of the temperature, and we would be back to the case of infinite thermal conductivity again.

Accordingly we must modify the above picture of thermal energy transport to allow for collisional scattering processes. Let us adopt the phonon-gas approach and consider the lattice replaced by a volume full of phonons of different frequencies. We picture the process of thermal conduction to be one where the temperature gradient at the hot face of the solid continually increases the total momentum of the phonon gas in the direction of heat flow (just as, earlier in this chapter, an externally applied electric field increased the momentum of the conduction electrons), and the scattering process removes this excess momentum as the phonons move from the hot face of the solid to the cool face. Each scattering process 'resists' the flux of phonons and therefore contributes to the thermal resistivity W which is the reciprocal of the thermal conductivity. But what are the scattering mechanisms which scatter the phonons as they travel through the solid? The answer is much the same as in the electrical conduction and valence-electron gas thermal conduction processes that we have already considered. The carriers of thermal energy (the phonons) are scattered by

a. Phonon–electron scattering collisions
b. Phonon–phonon scattering collisions
c. Phonon scattering by impurities and lattice defects
d. Phonon scattering by the actual boundaries of the solid.

In our discussion of electrical conductivity and valence-electron gas thermal conductivity we found that electron–phonon scattering was by far the dominant scattering process at and around room temperature for all conductors and that impurity/lattice defect scattering or boundary scattering usually only became important at lower temperatures. A similar conclusion holds here, where phonon–phonon scattering is the dominant scattering mechanism that controls the *lattice* thermal conductivity of solids at and around room temperature ($T > \theta_D$) and the other processes of impurity/lattice defect or boundary scattering only become important at the lower temperatures $T \ll \theta_D$. To a good approximation these three scattering mechanisms are independent of each other and we get an equation for W analogous to Matthiesen's rule [see equations (6.38) and (6.57)].

By now, the thoughtful reader will be confused. We have categorically stated that phonons of different frequencies do not interact with each other in the Debye approximation describing the lattice and yet we have just stated that phonon–phonon scattering is the dominant 'resistance' to the transfer of thermal energy; clearly both cannot be correct! The resolution of this inconsistency requires a slight modification of Debye theory. It is easier to describe such a modification by considering the quantum oscillator of Section 5.2. In that section we discussed the harmonic oscillator and showed that such an oscillator obeyed Hooke's law, i.e. from equation (5.3)

Potential energy of a stretched spring obeying Hooke's law

$$= \frac{f}{2}(x - x_e)^2$$

Analogously to equation (5.9) we can calculate,

Average extension of the spring over a period of oscillation

$$= \overline{(x - x_e)}$$

$$= \frac{\int_0^{\frac{1}{\nu}} (x - x_e)\,dt}{\int_0^{\frac{1}{\nu}} dt}$$

$$= 0$$

i.e. the average extension of the oscillator is zero. If we were considering the Einstein approximation of the lattice where the lattice is made up of independent nuclei undergoing harmonic oscillations, then such an Einstein solid would not expand on heating, thus it would have zero thermal expansion. This conclusion of zero thermal expansion also holds for the Debye model of the lattice since this treats the motion of very small volume elements of the solid also as being simple harmonic. But it is known that solids do expand on heating. The answer to this confusion lies in stating that Hooke's law is not obeyed exactly for an oscillator—only approximately. To be more precise we have to consider the anharmonic oscillator and write

Potential energy of an anharmonic oscillator

$$= \frac{f}{2}(x - x_e)^2 - B(x - x_e)^3 - C(x - x_e)^4$$

Here we have written the potential energy as a power series in the extension $(x - x_e)$ which is of course the departure from the equilibrium separation (see Figure 5.1). For very small oscillations the B and C coefficients are small so that we can ignore the cubic and quartic terms so that the motion is simple harmonic and symmetrical about the equilibrium position. However, for larger extensions this symmetry of oscillation disappears. The cubic term represents the asymmetry of the anharmonic oscillations; it steepens the forces of repulsion and flattens those of attraction. The quartic term softens the anharmonic oscillations at large amplitude by reducing the energy necessary to produce an extension. The average extension of the oscillator over a period of one anharmonic oscillation is now not zero but in fact increases as the extension increases.

In terms of both Einstein and Debye theories of the solid, anharmonic oscillations increase the average separation of either the lattice nuclei (Einstein model) or of a small-volume elements of the solid (Debye model) as the temperature increases and thereby permits thermal expansion to occur. But what has this got to do with interacting phonons? To answer this question let us return to the Debye description of the solid as an elastic continuum with thermal elastic waves moving through it. Now in practise, elastic waves (not necessarily thermal) do not pass through a solid without interacting with each other. This interaction (wave scattering) arises because the speed of sound in an elastic solid depends upon the density of the solid and the modulus of rigidity according to

$$\text{Square of speed of sound} = \frac{\text{Rigidity modulus of solid}}{\text{Density of solid}}$$

If the elastic wave motion is simple harmonic, then as we have already remarked, if the density of any small-volume element of the solid is increased (due to compression of that volume element analogous to the compression of a spring) then the rigidity modulus is increased by exactly the same amount so that the speed of sound is unchanged. However, if the wave motion is anharmonic then the change in density of a small-volume element when compressed is not exactly matched by a change in the rigidity modulus and so the speed of propagation of the waves, the speed of sound, changes as the waves pass through the solid. This means that subsequent waves are moving into regions of the solid where the density and rigidity modulus are changing from point to point and this results in these waves being scattered.

In phonon language, the phonons would travel without 'resistance to motion' throughout the entire volume of the solid if the thermal wave motion were one of simple harmonic oscillations. If, however, the motion is one of anharmonic oscillations (even very midly so that the anharmonicity can be regarded as only a perturbation on the harmonic oscillation) then a phonon will move into volume elements of the solid whose density and elastic properties are changing from point to point and in this case the phonons will be scattered and this is how we obtain phonon–phonon scattering interactions.

With this slight modification of Debye theory, let us return to our picture of thermal-energy transport. Let us try to use as much of the reasoning given in the electrical conductivity analysis and also the valence-electron gas thermal conductivity analysis to the problem of the lattice thermal-conductivity. We know that when material particles (those of non-zero rest mass) collide, then any excess of one energy over the other is shared so that after a sequence of collisions the excess energy of any particular particle above the average energy of the remainder of the particles is shared out so that eventually all particles in the container share what was originally the excess energy as they move along

Transport Properties of Solids

their *random* paths in space. However, consider the collision of two phonons; we know that analogously to photons, when two phonons collide they can 'coalesce' and form a single phonon that conserves the momentum and energy of the original collision partners; conversely a single phonon can split up into two phonons once again with momentum and energy conserved. We should note that in fact phonons do not carry real momentum—they only possess crystal or pseudomomentum, thus in a lattice they behave as if they possess momentum when they are involved in a scattering collision. This apparent momentum is in some ways analogous to the idea that an electron moving in an energy band behaves as if it had a variable, i.e. effective, mass. However, although energy and momentum are conserved in the collision between two phonons, nevertheless the net flux of thermal energy in the solid is not decreased or scattered in random directions. This is an n-process phonon–phonon scattering collision such that n-processes cause the scattering of phonons but such scattering merely redistributes the thermal energy into different phonon frequencies without actually altering the total phonon flux. If then the total phonon flux is unaltered, such n-processes still cannot provide the required 'resistance to motion' in the lattice thermal-conductivity problem. What sort of process do we require? The answer may be deduced from a consideration of the electrical conductivity problem in Section 6.3 where we showed that in electron–phonon scattering collisions, the electron loses only a very small amount of energy but changes its momentum. In the lattice thermal-conductivity problem we therefore require a phonon–phonon scattering collision where energy is conserved but momentum is not. This is a u-process (umklapp process) which we can picture as one where two phonons of high frequency both travelling in essentially the same direction (from the hot face of the solid to the cool face) coalesce; we might expect the result to be the creation of a phonon of even higher frequency moving in the same direction. However, in a u-process the net effect is equivalent to the creation of a phonon of lower frequency moving in the opposite direction! We shall not attempt to explain the u-process; this is a problem in quantum mechanics and not appropriate to a textbook on thermodynamics. u-processes can only occur if anharmonic oscillations of the lattice are considered and they do not occur in the harmonic oscillator Einstein or Debye models. Of course, although momentum is not conserved between the three phonons concerned in a u-process it is conserved generally as some is transformed back to the lattice. The rate at which u-processes occur depends upon the temperature of the solid and in fact a minimum phonon energy is required for a u-process to occur at all. This means that u-processes disappear altogether at low temperatures; we shall return to this point later. We can now give a complete picture of the process of lattice thermal conduction: n-processes, although not contributing to the thermal conductivity, directly result in low-frequency phonons combining with each other to form high-frequency phonons. These high-frequency phonons then suffer u-process phonon–phonon scattering collisions which result in a finite (as opposed to zero) thermal resistivity. u-processes also of course establish thermal equilibrium amongst the phonons which enables us to meaningfully refer to the 'temperature' of any small-volume element of the solid and its variation throughout the volume of the solid.

Now that we have provided some sort of picture of the phonon–phonon scattering mechanism let us return to our calculation of the lattice thermal conductivity. If we wish to use the formula given by equation (6.42).

$$\lambda = \frac{1}{3} \frac{C_V}{V} \overline{v_{\text{th}}} l$$

we have to:

a. Modify the value of C_V given by Debye theory in equations (5.23), (5.27) and (5.30) to allow for the slight anharmonicity of the lattice oscillations because these equations were based on the lattice harmonic oscillator approximation.
b. Identify both $\overline{v_{\text{th}}}$ and l.

Although we shall not prove it, a simple analysis shows that the difference between the constant-volume heat capacity for harmonic oscillations of the lattice and for anharmonic oscillations can be disregarded, certainly within the limitations of the approximate analysis that we have given; we can continue to use the Debye equations. The identification of $\overline{v_{\text{th}}}$ is not difficult; this was originally introduced as the root mean square thermal speed of the carriers participating in the transport process (see Section 6.5). In the present case the carriers are the phonons and as we have stated in Section 5.5 these all travel through the solid at the speed of sound so that there is *not* a distribution of phonon speeds [as there is for material particles in a thermally perfect gas; see for example the Maxwellian distribution of speeds given in equation (2.68)]. Phonons all travel through the solid at the sound speed v_p which is practically constant, insensitive to changes in temperature of the solid having a typical numerical value of 5×10^3 m/s for most solids. There is a similar behaviour here with the electrical conductivity case where we found that the conduction electrons all moved essentially with the same temperature-independent speed—the Fermi speed.

Hence, just as in the electrical conductivity case, we find that the lattice thermal conductivity is determined by the phonon mean free path l which in turn is determined by u-processes. Without getting too involved in a detailed quantum analysis, we can say analogously to equation (6.32) that the phonon mean free path is

$$l \propto \frac{1}{\text{number of phonons per unit volume}}$$

Transport Properties of Solids

Let us first consider the high-temperature limit. From equation (5.38) we have for $T > \theta_D$

$$\text{number of phonons per unit volume} = \frac{N_p}{V}$$

$$= \frac{9NT}{2V\theta_D}$$

Also from equation (5.27)

$$C_V = 3kN$$

Hence, applying equation (6.42) to the lattice thermal conductivity

$$\lambda_L \propto \frac{1}{3}\frac{3kN}{V} \cdot v_p \cdot \frac{2\theta_D \cdot V}{9NT}$$

i.e.

$$\lambda_L \propto \frac{v_p \cdot \theta_D}{T} \tag{6.58}$$

Provided that v_p, θ_D are sensibly independent of temperature then

$$\lambda_L \propto \frac{1}{T} \tag{6.59}$$

Experiment shows that equation (6.59) holds for temperatures down to about $\frac{\theta_D}{10}$. We therefore obtain the surprising result that the lattice thermal conductivity *decreases* as the temperature increases. We might have expected that since the thermal conductivity depends upon the number of phonons—and that these phonon numbers increase with increasing temperature [see equation (5.38)] then the lattice thermal conductivity would have increased with increasing temperature. The opposite and correct explanation is that although the phonon numbers certainly do increase as the temperature increases, they become so abundant that they 'get in each other's way', that is, they shorten their own mean free path and thereby *lower* the thermal conductivity.

Let us now consider the low-temperature limit, $T \ll \theta_D$. From equation (5.39) we have

$$\text{number of phonons per unit volume} = 21 \cdot 64 \frac{N}{V}\left(\frac{T}{\theta_D}\right)^3$$

Also, from equation (5.30) for $T < \dfrac{\theta_D}{10}$

$$C_V = \frac{12\pi^4}{5} Nk \left(\frac{T}{\theta_D}\right)^3$$

Hence the lattice thermal conductivity is

$$\lambda_L \propto \frac{1}{3} \frac{12\pi^4}{5} \cdot \frac{N}{V} k \left(\frac{T}{\theta_D}\right)^3 v_p \frac{1}{21\cdot 64 \dfrac{N}{V} \left(\dfrac{T}{\theta_D}\right)^3}$$

i.e.

$$\lambda_L \propto v_p \tag{6.60}$$

If v_p is independent of the temperature, as we have stated it is, then

$$\lambda_L = \text{constant} \tag{6.61}$$

Therefore, as the temperature falls towards absolute zero from, say, $\dfrac{\theta_D}{20}$ the lattice thermal conductivity should remain constant.

This limiting behaviour of the lattice thermal conductivity is *not* found in practise. The reason is that at these low temperatures the phonon density has fallen to such a low value that u-process phonon–phonon scattering collisions cease to be important in limiting the phonon mean-free path. We have already referred to this above when we saw that u-processes require a certain minimum phonon energy and that at low temperatures there are just not sufficient high-energy phonons present to render u-processes effective.

What then determines the low-temperature behaviour of the lattice thermal conductivity? To answer this, let us consider the intermediate temperature region below the Debye temperature. As the temperature of the solid falls below the Debye temperature, the lattice thermal conductivity rises—this is because, although the absolute number of phonons is decreasing, the phonon–phonon mean free path is increasing. The increase in the phonon–phonon mean free path is greater than the reduction in the number of phonons so that the lattice thermal conductivity increases. This increase, occuring as the temperature falls, is very rapid; for example, between $\dfrac{\theta_D}{10}$ and $\dfrac{\theta_D}{20}$ it can be shown to be proportional to $\exp \dfrac{\theta_D}{2T}$ provided that the solid is chemically pure and free from lattice defects. Eventually as the temperature falls towards about $\dfrac{\theta_D}{20}$, u-processes cease to become dominant so that the lattice thermal conductivity

Transport Properties of Solids

reaches a maximum at about $\frac{\theta_D}{20}$. Thereafter, as the temperature falls even lower the lattice thermal conductivity falls because the phonon mean free path becomes limited, not by u-processes but by other scattering processes. For a chemically pure solid free of lattice defects, the phonons are scattered by the actual boundaries of the solid, in other words they effectively travel the entire length of the solid without scattering and then suffer scattering at the edges of the solid. This behaviour is completely analogous to the electrical conductivity of thin metallic films and the valence-electron gas thermal conductivity of pure conductors at low temperatures. Clearly, in the present case, the phonon mean free path becomes constant with a value equal to the distance between the hot and cool faces of the solid. Under such circumstances the lattice thermal conductivity is

$$\lambda_L = \frac{1}{3} \cdot \frac{12\pi^4}{5} \cdot \frac{Nk}{V} \left(\frac{T}{\theta_D}\right)^3 v_p \cdot l_{\text{solid}}$$

i.e.

$$\lambda_L \propto T^3 \tag{6.62}$$

so that for temperatures $< \frac{\theta_D}{20}$ the lattice thermal conductivity falls to zero as T^3—provided that the solid is chemically pure and free from lattice defects so that boundary scattering dictates the phonon mean free path.

A completely rigorous theory that accurately predicts the numerical value and complete temperature dependence of the lattice thermal conductivity is not yet available and it is one of the current problems of solid-state theory. The best analyses currently available show that

$$\lambda_L = \lambda_0 f\left(\frac{\theta_D}{T}\right) \tag{6.63}$$

where

$$f\left(\frac{\theta_D}{T}\right) = \frac{\theta_D}{T} \quad \text{for} \quad T > \theta_D$$

and

$$f\left(\frac{\theta_D}{T}\right) = \left(\frac{T}{\theta_D}\right)^3 \exp\left(\frac{\theta_D}{bT}\right) \quad \text{for} \quad T < \frac{\theta_D}{20}$$

where b is a constant, unfortunately not given accurately by the theory!

If the solid contains chemical impurities and/or lattice defects then the behaviour of the lattice thermal conductivity as a function of temperature is different. The high-temperature behaviour is still given by equation (6.59) but the exponential increase of the lattice thermal conductivity with decreasing temperature (on the high-temperature side of the maximum) does not occur. It is found that the maximum in the lattice thermal conductivity is considerably reduced and that it increases as $T^{\frac{3}{2}}$ as the temperature decreases. Figure 6.4

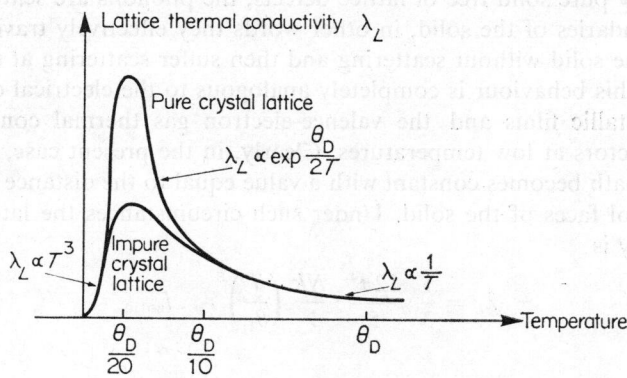

Figure 6.4. Schematic plot of the variation of lattice thermal conductivity λ_L of a pure and of an impure crystal as a function of temperature (the temperature scale is non-linear).

shows a schematic representation of the temperature dependence of the lattice thermal conductivity of a typical solid, both pure and impure.

6.8 The total thermal conductivity of a solid

In general the total thermal conductivity λ_T is the sum of the valence-electron gas and lattice contributions since the scattering mechanisms act essentially independently of each other, i.e.

$$\lambda_T = \lambda_e + \lambda_L \qquad (6.64)$$

Clearly, a question immediately arises as to whether the valence electrons or phonons carry the greater part of the thermal energy. The answer depends entirely on whether the solid is a metal, insulator or semiconductor (all of which we shall define in Chapter 7). We get some clue to the answer by noting that at room temperature pure metals are found to have thermal conductivities λ_T up to perhaps one hundred (or even in certain cases one thousand) times greater than those of insulators; for example a metallic conductor such as silver has a room-temperature value of total thermal conductivity of 420 W (m deg C)$^{-1}$ whilst that of a poor electrical conductor like quartz is 13 W (m deg C)$^{-1}$, i.e. a factor of about 30 lower. However, as we have already noted, their electrical conductivities differ by a factor about 10^{22} or even higher.

Transport Properties of Solids

This enormous difference in behaviour between the thermal and electrical conductivities must be related to the fact that in the process of electrical conduction, only conduction electrons participate since phonons carry no charge. However, for thermal conduction *both* conduction electrons and phonons are potentially able to participate so that the thermal conductivity (although affected by the conduction electrons) is maintained by phonon transport in insulators.

Let us therefore consider the total thermal conductivity of a metal. In most metals it is found that

$$\lambda_e > 10\lambda_L \qquad (6.65)$$

so that at ordinary temperatures we need not consider the lattice contribution to the total thermal conductivity at all. In the laboratory it is not possible in general to measure the thermal conductivity of a solid with an accuracy greater than 1% so that it is not usually possible to determine the lattice contribution to the total thermal conductivity of a metal with any great accuracy. The reason why the lattice contribution is so small is that phonons are strongly scattered by the large number of conduction electrons in a metal. This results in a very small phonon mean free path so that the lattice thermal conductivity is small [see equation (6.42)]. Hence for a metal

$$\lambda_T \sim \lambda_e \qquad (6.66)$$

Of course the phonons, although not contributing to the total thermal conductivity, do participate indirectly, that is, they represent scattering centres for the 'heat carrying' conduction electrons so that electron–phonon scattering of the conduction electrons determines the valence-electron gas thermal conductivity and therefore the total thermal conductivity, as we have already concluded in Section 6.4.

The result indicated by equation (6.66) that the total thermal conductivity is equal to the valence-electron gas thermal conductivity is valid for most metals. However, for metals like Bi and Sb the conduction electron concentration is low and we must, in these rather exceptional cases, consider the contribution of the lattice to the total thermal conductivity. Another case where the lattice thermal conductivity of a metal cannot be ignored is that of very impure metals. We have already noted in Section 6.6 that in such impure metals conduction electron scattering by impurities tends to dominate the valence-electron gas thermal conductivity at *all* temperatures so that in this case the lattice thermal conductivity must be considered. In fact, this is sometimes used in experimental determinations of the lattice thermal conductivity which, as we have just remarked, is not accurately experimentally determinable for metals; impurities are added that reduce the valence-electron gas thermal conductivity but do not significantly alter the lattice thermal conductivity since such impurities tend to scatter the high-frequency phonons only and leave the low-frequency phonons essentially unaffected. Accordingly, low-temperatures studies of very impure metals give quite accurate values of the lattice thermal conductivity.

The most important class of substances where equation (6.66) does not hold and where we must consider the contributions of both the valence electrons and the lattice to the total thermal conductivity are the semiconductors and we shall return to a discussion of these in Chapter 7.

For insulators, of course, since these have effectively no conduction electrons at all, the total thermal conductivity is

$$\lambda_T \sim \lambda_L \qquad (6.67)$$

where asymptotic values of the lattice thermal conductivity have already been given by equations (6.59) and (6.62).

All of the considerations so far have been for rather ideal single crystalline solids. For polycrystalline dielectrics at very low temperatures the phonons are scattered at the boundaries of crystallites rather than at the geometrical boundaries of the solid, and the phonon mean free path is strictly limited and does not vary as strongly with temperature as in the case of the single-crystalline solid. The total thermal conductivities of commercially important dielectrics such as glass, perspex, nylon, etc., are much smaller (by a factor that can be as much as 10^3) than the ideal single-crystal dielectric; also, their temperature dependence is such that the lattice thermal conductivity decreases more or less monotonically from high temperature to low and there is no maximum. Some typical values of the total thermal conductivity of non-metals at room temperature are given in Table 6.3.

Table 6.3 Total thermal conductivity λ_T of a selection of non-metals; these are essentially the lattice thermal conductivities.

Substance	λ_T W(m deg C)$^{-1}$
Quartz	12·5
NaCl	7·0
Carbon	4·2
Pyrex (silica)	1·0
Mica	0·8

6.9 The Wiedemann–Franz–Lorenz law for electrical conductors

All of this chapter on transport processes has been concerned with the motion of the carriers of either negative electrical charge in the electrical conductivity

Transport Properties of Solids

problem and the valence-electron gas thermal conductivity problem or phonons in the lattice thermal conduction problem. It will now be obvious that the essence of the transport process problem has been a determination of the causes of 'resistance to motion' of these carriers as they move throughout the volume of the solid, in particular an evaluation of the mean free path l.

In the electrical conductivity case we have shown that the conduction electrons are impeded by scattering collisions with either phonons or impurity atoms/lattice defects or, in special cases, the boundaries of the solid. Let us summarize the equations that we have derived for the electrical resistivity. Equation (6.33) which arises because of conduction electron–phonon scattering is

$$\rho_i \propto T$$

for the temperature range $T > \theta_D$. However, if the temperature range is $\frac{\theta_D}{30} < T < \frac{\theta_D}{10}$ then this equation becomes (see Section 6.4)

$$\rho_i \propto T^5$$

which also applies to electron–phonon scattering. If however, the collisional mechanism is conduction electron-impurity atom scattering then we had equation (6.37) for the residual resistivity, viz.

$$\rho = \rho_0 = \text{constant}$$

which holds for the temperature range $T < \frac{\theta_D}{30}$.

We combined these equations and wrote equation (6.38) for the total electrical resistivity as

$$\rho_T = \rho_i(T) + \rho_0$$

In the valence-electron gas thermal conductivity analysis we showed that the conduction electrons were likewise scattered either by collisions with phonons or with impurity atoms/lattice defects or with the boundaries of the solid. Summarizing these equations we found that

$$\lambda_e = \text{constant} \quad \text{for} \quad T > \theta_D$$

which is equation (6.48) and holds for electron–phonon scattering. If the temperature is $\frac{\theta_D}{30} < T < \frac{\theta_D}{10}$ then the appropriate equation for electron–phonon scattering is [see equation (6.50)]

$$\lambda_e \propto T^{-2}$$

If, however, we are considering electron-impurity atom scattering then we obtained equation (6.55) which was

$$\lambda_e \propto T \quad \text{for} \quad T < \frac{\theta_D}{30}$$

We combined these equations and wrote equation (6.57) for the total valence-electron gas thermal conductivity

$$\left(\frac{1}{\lambda_e}\right)_T = \alpha T^2 + \frac{\beta}{T} \quad \text{for} \quad T < \frac{\theta_D}{20}, \text{ say.}$$

Consider the product of total electrical resistivity times total valence-electron gas thermal conductivity

$$\rho_T(\lambda_e)_T = \{\rho_0 + \rho_i(T)\} \cdot \{(\lambda_e)_T\} \tag{6.68}$$

The limiting forms of this product are:
High-temperature limit for dominant electron–phonon scattering processes

$$\rho_T \sim \rho_i(T) = C_1 T \quad \text{for} \quad T > \theta_D \text{ from equation (6.33)}$$

since

$$\rho_0 \ll \rho_i(T)$$

$$\lambda_e = C_2 \quad \text{for} \quad T > \theta_D \text{ from equation (6.48)}$$

where C_1 and C_2 are constants

$\therefore$

$$\rho_T(\lambda_e)_T = C_1 C_2 T$$

i.e.

$$\frac{\rho_T(\lambda_e)_T}{T} = \text{constant} \tag{6.69}$$

We see that the right-hand side is independent of temperature.

Turning to the low-temperature limit for dominant electron–phonon scattering processes we have

$$\rho_T = \rho_0 + C_3 T^5 \quad \text{for} \quad T < \frac{\theta_D}{10}$$

$$(\lambda_e)_T = \frac{1}{\alpha T^2 + \dfrac{\beta}{T}} \quad \text{for} \quad T < \frac{\theta_D}{20}$$

$\therefore$

$$\rho_T(\lambda_e)_T = \frac{\rho_0 + C_3 T^5}{\alpha T^2 + \dfrac{\beta}{T}} \quad \text{for} \quad T < \frac{\theta_D}{20}$$

For very pure specimens

$$\rho_0 \sim 0 \quad \beta \sim 0$$

∴

$$\rho_T(\lambda_e)_T \simeq \frac{C_3}{\alpha} \cdot T^3 \tag{6.70}$$

We see that the right-hand side is *not* independent of temperature. However, if the specimen is not pure then the low-temperature limit for dominant electron-impurity atom scattering is found thus

$$\rho_T \sim \rho_0 \quad \text{for} \quad T < \frac{\theta_D}{30}$$

$$(\lambda_e)_T \sim \frac{T}{\beta} \quad \text{for} \quad T < \frac{\theta_D}{20}$$

∴

$$\rho_T(\lambda_e)_T = \frac{\rho_0}{\beta} \cdot T \quad \text{for} \quad T < \frac{\theta_D}{30}$$

∴

$$\frac{\rho_T(\lambda_e)_T}{T} = \text{constant for} \quad T < \frac{\theta_D}{30} \tag{6.71}$$

We therefore see that the ratio $\frac{\rho_T(\lambda_e)_T}{T}$ is a constant for the two following cases:

a. For all conductors at temperatures greater than the Debye temperature when electron–phonon scattering collisions are responsible for the 'resistance to motion' of the conduction electrons.

b. For impure conductors at temperatures less than about $\frac{\theta_D}{30}$ such that electron-impurity atom scattering collisions cause the 'resistance to motion' of the conduction electrons.

The ratio $\frac{\rho_T(\lambda_e)_T}{T}$ is called the Lorenz number L of the conductor whilst the equation

$$\rho_T(\lambda_e)_T = LT \tag{6.72}$$

is called the Weidemann–Franz–Lorenz law. This law, which as we have shown holds for 'high' ($T > \theta_D$) and 'low' ($T \ll \theta_D$) temperatures is of considerable theoretical importance because it gives information on conduction electron scattering processes in a conductor. It is also of great practical importance since it is usually possible to make accurate laboratory experimental measurements of electrical resistivity and from such measurements together with a knowledge of the Lorenz number and temperature the valence-electron gas thermal conductivity can be deduced thereby avoiding the necessity of making what are difficult experimental measurements of this transport coefficient.

If we combine the 'impurity' equations for the residual resistivity and the impurity atom component of the valence-electron gas thermal conductivity then we get

$$\frac{\rho_T(\lambda_e)_T}{T} = \frac{\rho_0}{\beta} = L$$

i.e.

$$\beta = \frac{\rho_0}{L} \tag{6.73}$$

We can therefore obtain values of the impurity coefficient β which we introduced in equation (6.56) using known values of the residual resistivity which have been given in Table 6.1.

The discussion so far has been confined to the 'high' and 'low'-temperature regions; what happens in the intermediate, region say $\frac{\theta_D}{30} < T < \theta_D$? Because our analysis of transport processes has been only semiquantitative at best, we cannot prove in this book the fact that the Wiedemann–Franz–Lorenz law does *not* hold at intermediate temperatures. However, let us see what general comments can be made. Equations (6.17) and (6.24) give

$$\rho_i = \frac{m_e^* v_{\text{Fermi}}}{n_e \cdot e^2} \left(\frac{1}{l_{\text{Fermi}}} \right)_\rho$$

whilst equations (6.43) and (6.24) give

$$\lambda_e = \frac{\pi^2}{3} \cdot \frac{k^2 T}{m_e^*} \cdot n_e (l_{\text{Fermi}})_\lambda$$

Hence

$$\frac{\rho_i \lambda_e}{T} = \frac{\pi^2}{3} \cdot \left(\frac{k}{e} \right)^2 \frac{(l_{\text{Fermi}})_\lambda}{(l_{\text{Fermi}})_\rho}$$

$$= 2.45 \times 10^{-8} \frac{(l_{\text{Fermi}})_\lambda}{(l_{\text{Fermi}})_\rho} \, \text{W} \, \Omega \, \text{deg} \, \text{C}^{-2}$$

If the two Fermi mean free paths are the same then the value of the Lorenz number should be

$$L = 2.45 \times 10^{-8} \, \text{W} \, \Omega \, \text{deg} \, \text{C}^{-2}$$

Transport Properties of Solids

For many conductors experimentally determined Lorenz numbers come fairly close to this figure when evaluated for room-temperature conditions *or* for temperatures below about 5°K. Table 6.4 gives some values of the experimentally determined ratio $\dfrac{\rho_T(\lambda_e)_T}{T}$ for some conductors at 0°C.

Table 6.4 Values of the experimentally determined ratio $\dfrac{\rho_T(\lambda_e)_T}{T}$ for a selection of metals at 0°C.

Conductor	$\dfrac{\rho_T(\lambda_e)_T}{T}$ W Ω deg C^{-2} × 10^8
Cu	2·23
Ag	2·31
Au	2·35
Zn	2·31
Cd	2·42
Pb	2·45
Bi	4·26

Why does the law fail in the intermediate temperature range? A full and detailed answer to this question cannot be given here since it requires a deeper quantal discussion of transport processes than we have given. The essence of the answer is that the ratio $\dfrac{(l_{\text{Fermi}})_\lambda}{(l_{\text{Fermi}})_\rho}$ does not always have the value of unity. The reason for this behaviour is that in the intermediate temperature region $\dfrac{\theta_D}{30} < T < \theta_D$ there is a fundamental difference between the conduction electron–phonon scattering process in the case of thermal conduction compared with the case of electrical conduction, *provided* we are in that region where electron–phonon scattering collisions are still dominant in providing 'resistance to motion' of the conduction electrons. If this condition is satisfied (and it is *not* satisfied as we have seen for impure conductors at very low temperatures equation (6.71) showing that the Wiedemann–Franz–Lorenz law *is* obeyed) then most of the phonons are those of low frequency because the temperature of the conductor is quite low. For the electrical resistivity case in an electron–phonon collision the transfer of *momentum* is the important issue. In Section 6.3 we showed that the change of energy in such a collision was negligible and that such change of momentum was quite substantial—provided that the solid was at a temperature greater that the Debye temperature. However, in the intermediate temperature range all of the phonons have low frequency and all of the phonons have low momentum. When an electron collides with a low-momentum phonon then there is only a very small exchange of momentum, thus the electron is deflected through only a very small angle. Clearly if the electron is to lose a significant amount of momentum (as it must do in the electrical resistivity problem) then it will need to make many such small-angle scattering collisions. Therefore low-frequency phonons are poor scatterers of conduction electrons. However, in the thermal resistivity case we must recall that there is no net flow of conduction electrons. At any point within the conductor there will be conduction electrons flowing down the temperature gradient which are slightly 'hotter' than the *equal number* of conduction electrons flowing in the opposite direction. Once again there will be small-angle electron–phonon scattering but the *energy* exchanged will be about kT and it is energy transfer that is important; this is quite sufficient to 'cool' an 'above-average' energetic conduction electron below the average energy. Hence, low-frequency phonons are good scatterers of thermal-energy conduction electrons. Consequently, in the intermediate temperature range where electron–phonon collisions dominate the transport process the 'resistance to motion' is very different, i.e. we have

$$(l_{\text{Fermi}})_\rho > (l_{\text{Fermi}})_\lambda$$

and this of course is demonstrated in the results already quoted above, viz.

$$\rho_i(T) \propto T^5$$
$$W_e(T) \propto T^2$$

This explains why we get the behaviour given in equation (6.70) where the Wiedemann–Franz–Lorenz law is not obeyed.

Problems to Chapter 6

1. A block of copper at 295°K is in the form of a cube of side 5×10^{-3} m and has a measured electrical resistance of 3·4 micro-ohms. If the atomic weight of copper is 63·54 kg(kg mole)$^{-1}$ and its density at 295°K is 8.94×10^3 kg/m^3 find
   a. The valence electron density in copper assuming unit valency.
   b. The valence electron mobility which is given by $(\text{pen}_e)^{-1}$ where ρ is the electrical resistivity and n_e the valence electron concentration.
   c. The electron scattering collisional relaxation time for conduction electrons.
   d. The Fermi speed v_{Fermi} and the Fermi scattering length l_{Fermi}.
   e. The drift velocity of the conduction electrons when an electric field of 10^4 V/m is applied to the copper block.

Answers: 8.47×10^{28} m^{-3}; 4.34×10^{-3} m^2/Vs; 2.47×10^{-14} s; 1.57×10^6 ms^{-1}, 3.89×10^{-8} m; 43·4 ms^{-1}.

Transport Properties of Solids

2. An important aspect of transistor operation is temperature control of the junction since its electrical characteristics are sensitive to temperature. One way of keeping the junction cool is to mount the transistor on a heat sink which is usually a block of metal with a high free-electron gas thermal conductivity.

A power transistor dissipating 35 watts is mounted on a circular silver heat sink of radius 0·5 cm and thickness 1 cm. Calculate the temperature difference between the hot and cool faces of the heat sink.

Answer: 10·6 deg C.

CHAPTER SEVEN

The Quasifree Thermally Perfect Electron Gas in a Semiconductor

7.1 Electron energy bands in a solid—the valence and conduction bands

The valence-electron gas theory discussed in Chapter 4 gave us considerable information about some important thermodynamic properties of metals whilst the transport theory developed in Chapter 6 gave us information about the electrical and thermal conductivities of solids in general.

Now every solid, be it a metal or an insulator, contains electrons, and since these two classes of solids are found to have totally different electrical and thermal properties we must ask ourselves what proportion of these electrons in the solid are available for thermodynamic (thermal) and electrical processes. The answer to this question requires us to take a closer look at the valence-electron gas model of Chapter 4 and see if it is sufficiently realistic to explain why there are differences between metals and insulators. In Chapter 4 we used the Sommerfeld model of the solid and assumed that the valence electrons moved in a flat-bottomed potential well [see Figure 4.1(b)], thus we completely disregarded the periodic electrostatic potential that must exist in the solid due to the regularly spaced lattice nuclei. What we did was to replace these regularly spaced positive charges (which must be there to keep the solid electrically neutral) by a fixed uniform background of positive charge in which the valence electrons were quite free to move. We commented that this seemed a fair approximation because the valence electrons have a very small mass and move through the entire volume of the solid with speeds greatly in excess of the speed of the nuclei as they performed their thermal oscillations about their equilibrium positions. Therefore since the nuclei are essentially at rest compared with the highly mobile valence electrons, we can replace them by a fixed uniform background of positive charge. There is further justification for this smoothing out of the positive potential. The strong Coulomb field of each lattice nucleus is not the simple inverse power of an isolated charge; rather, each nuclear charge is 'screened' from the field of the other nuclear charges by the valence-

The Quasifree Thermally Perfect Electron Gas

electron gas (Debye–Hückel screening). This means that the potential seen by any valence electron moving through the solid is not as variable in space as it would be if the nuclei had possessed pure unscreened Coulomb potentials.

Nevertheless, considerations of screening alone will not enable us to alter the valence-electron gas model sufficiently to explain the great differences between metals and insulators or the existence of that very important class of materials called semiconductors. What we need to do is to abandon the oversimplifications of the Sommerfeld model and incorporate the effect of the lattice structure into our calculations, that is, solve the Schrödinger wave equation not in the flat-bottomed potential well of the Sommerfeld model according to Figure 4.1(b) but in the periodic potential of Figure 4.1(a). We shall not give the quantum theory of this calculation since it is inappropriate for a textbook on thermodynamics. It is known as the Bloch model of a solid and is available in all of the standard texts on solid-state theory. Here we shall merely quote the results that we require. Bloch translational energy states bear some resemblance to the translational energy states of the thermally perfect electron gas (Sommerfeld model) shown in Figure 4.2—but with one very profound difference. The Bloch wave functions do *not* allow the valence electrons to have certain translational energies; hence we obtain a distribution of allowed, very narrowly spaced quantized translational energy levels (a 'band' of energy levels), then an energy gap where no valence electrons can remain in any stable allowed translational energy state, followed by another band of allowed translational energy levels. Figure 7.1(a) shows such a set of translational energy levels and the energy gap whilst Figure 7.1(b) shows a simplification of this set of allowed and forbidden translational energy-state bands that we shall use as the model for analysis in this chapter. The band containing the valence electrons is still called the valence band; above the valence band is the energy gap and above that there lies the conduction band. Within the allowed translational energy bands (which are given by the Bloch solutions of the Schrödinger wave equation) the electron translational energy states and levels are very similar to those considered in the valence-electron gas (Sommerfeld) model of Chapter 4. In fact we can think of the valence-electron gas model as giving a single band of allowed translational energy levels with no energy gap—we return to this below.

Thus far we have considered the allowed and forbidden valence electron translational energy states but we have not stated whether any particular translational energy state in a band is actually occupied by an electron. We shall use statistical thermodynamics to give us algebraic answers to this question of energy-state occupation. Let us anticipate the detailed answers and state that our analysis shows that the valence band need not be full of electrons (in the Pauli sense); it may be only partially filled so that there will be empty allowed electron translational energy states at the *top* edge of the valence band;

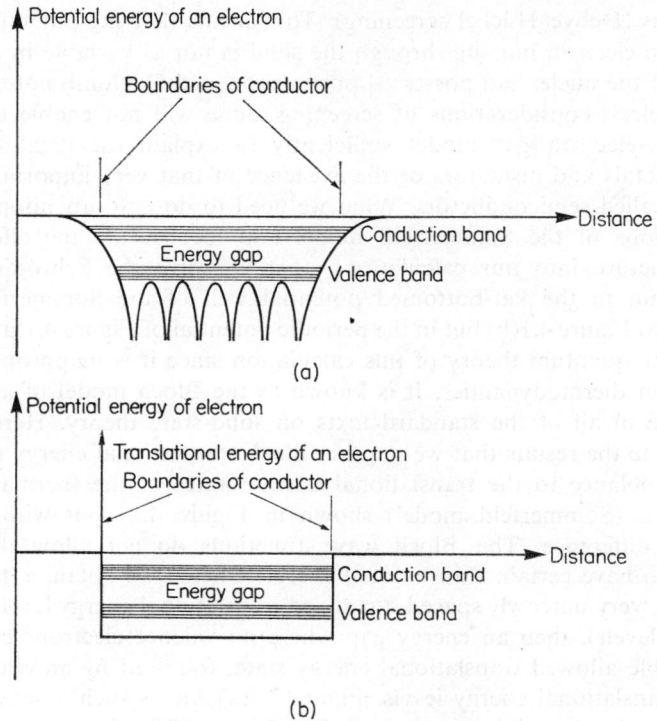

Figure 7.1(a). The actual distribution of potential energy of an electron as a function of its distance from the boundary of the conductor. The potential energy rises and falls periodically throughout the volume of the conductor. The valence electron is completely free to move in the valence band, or in the conduction band but not in the energy gap. This figure should be compared with Figure 4.1(a). (b) The simplified distribution of potential energy of an electron as a functional of its distance from the boundary of the conductor. The potential energy is constant throughout the entire volume of the conductor and forms a potential well of infinitely steep sides. Within this potential well there is a 'ladder' of allowed translational energy levels that form the valence band, separated by the energy gap from a higher lying 'ladder' of allowed translational energy levels that form the conduction band. This figure should be compared with Figure 4.1(b) and 4.2.

the conduction band will be empty of electrons if the valence band is only partially full. However, if the valence band is completely full then the conduction band may be empty or partially full. In fact, a partially filled valence or conduction band may have only a 'few' electrons in it, that is, considerably fewer electrons than there are allowed translational energy states, or it may be either approximately half full or 'essentially' full, with only a 'few' empty translational energy states remaining within it.

The Quasifree Thermally Perfect Electron Gas

The essence of the problem of electrical and thermal conduction *in a semiconductor* is the ability of a significant number of electrons to cross the energy gap from the valence band to the conduction band. Let us see why this is. Let us consider initially the problem of electrical conduction by supposing that the semiconductor is subjected to an externally applied electric field. In Chapter 6 we considered electrical conduction in a metal and showed in Section 6.3 (and also towards the end of Section 4.2) that if electrons are to be allowed to gain translational energy and momentum from the applied electric field then there must be empty allowed electron translational energy states for these electrons to move into. If we apply this consideration to the semiconductor then we see that electrical current conduction will obviously not occur in an empty allowed energy band (since there are no electrons to conduct) or in a completely *full* band (since there are no empty translational energy states available for the electrons to move into). Hence, current conduction is only possible in *partially filled bands* and such current-conduction properties will change if the number of electrons in a partially filled band changes; this will happen if valence electrons cross the energy gap. In this book we shall consider only one process for giving electrons sufficient energy to cross the energy gap from the valence band to the conduction band, namely, a gain of translational energy from an externally applied high-temperature source. This is *thermal* excitation of electrons across the energy gap due to the electrons gaining energy from the thermal energy of the oscillating lattice.

Before we discuss these matters quantitatively, we can now explain the difference between the electrical properties of various solids. Metals have a partially filled energy band due either to an overlap of the valence and conduction bands (i.e. no energy gap—noble metals), or the fact that the valence band is only partially filled (alkali metals). Insulators have an essentially full valence band separated by a large energy gap from an essentially empty conduction band; hence current conduction in an insulator is very unlikely. Semiconductors are essentially insulators, that is, they have an energy gap but it is of such small width that electrons can readily cross the gap if thermally excited. Now in the previous sentences we have been rather vague in using the word 'essentially'. This is because a solid which has a completely filled valence band and a completely empty conduction band will be a true insulator—but only at absolute zero. At all non-zero temperatures some electrons from the valence band *may* have sufficient energy to be thermally excited into the empty conduction band so that solely by virtue of temperature alone, the insulator has become a possible conductor. Clearly how effective a conductor it has become will depend upon the number of valence electrons thermally excited into the conduction band which in turn, as our statistical thermodynamic analysis will show, depends upon the width of the energy gap and the temperature.

We therefore conclude that the distinction between an insulator and a semiconductor is not a clear one, that is, we can never just talk about a completely full valence band or a completely empty conduction band since these only occur at the absolute zero, i.e. not only all insulators but also all semiconductors are truly insulators at the absolute zero; for example, diamond, silicon and germanium are all insulators at the absolute zero. However, at all temperatures above absolute zero there will be a small number (exceedingly small for insulators) of electrons present in the lowest energy states of what was an empty conduction band so that we can conclude that insulators are intrinsic semiconductors at temperatures above absolute zero! Consequently we shall not attempt a formal definition of a semiconductor. Perhaps the best way of regarding them is to say that, generally speaking, the electrical conductivities of semiconductors at room temperature lie within the range 10^{-7} to 10^4 $(\Omega \text{ m})^{-1}$ which is intermediate between metals whose electrical conductivities lie between 10^4 to 10^7 $(\Omega \text{ m})^{-1}$ and insulators whose electrical conductivities lie within the range 10^{-7} to 10^{-15} $(\Omega \text{ m})^{-1}$. In passing we again note the enormous range in these values of electrical conductivity; they range between 10^{-15} to 10^7 $(\Omega \text{ m})^{-1}$ which is a factor of 10^{22} and at cryogenic temperatures this factor can increase to 10^{32}. In Chapter 6 we noted that the thermal conductivities of metals to insulators range over a factor of 10^3 or less and this dramatic difference in behaviour has been discussed in Section 6.8.

Therefore, the ability of any solid to conduct electricity depends upon:

a. Whether the valence band is essentially full or only partially filled.
b. Whether the conduction band is essentially empty or partially filled.
c. The presence (or absence) of an energy gap and its width.
d. The temperature, since electrons can gain thermal energy from the lattice and move from the valence band into the conduction band.

Of course if semiconductors were only what we have just discussed then they would be of value but not of the greatest technological importance. The reader is assumed to know that what has made semiconductors so important is that their electrical properties *at and around room temperature* are profoundly modified by the presence of exceedingly small amounts of impurities. In Section 6.4 we discussed the problems of the residual resistivity of a metal and showed how impurities in the metal determine the electrical resistivity—but only at very low temperatures, say, $T < 0.03\theta_D$; however, in the case of a semiconductor the electrical resistivity is dramatically affected *at and around room temperature* by impurities. Furthermore, in the metallic case we found that 'spectrographically pure' metals had about one atom of impurity in 10^5 host atoms. New technological techniques have been developed for adding impurities in closely controlled amounts (doping) to semiconductors and the purest refinement available at present gives a minimum of about one impurity atom in 10^{10} host atoms.

The Quasifree Thermally Perfect Electron Gas

Since there are about 10^{28} atoms per cubic metre in most solids we see that doping a semiconductor can involve the addition of exceedingly small amounts of impurities. The amount of doping depends upon the use that the semiconductor is to be put to; for example, in a germanium transistor the base doping may be about 10^{21} m^{-3} or even lower whilst for tunnel diodes and thermoelectric materials the doping may be as high as 10^{25} m^{-3} or even higher. Although there are very many extrinsic semiconductors that have been developed for specific purposes the two most important, technologically speaking, are silicon and germanium and we shall use either of these to illustrate the calculations made further within this chapter.

We have now distinguished between the intrinsic semiconductor which is a 'pure' solid of small energy gap and the extrinsic semiconductor in which the semiconductor material has had impurities deliberately added to it so as to introduce extra electron energy states *in the energy gap*. The reader may justifiably ask whether there is such a thing as an intrinsic semiconductor at all; after all, it is known that the addition of only one atom of impurity in 10^7 host atoms can turn an intrinsic semiconductor into an extrinsic one. The reason why we study intrinsicity in semiconductors is that in some cases intrinsic semiconductors *can* be made at room temperature (we shall later prove that germanium can but silicon cannot) and in addition, as we shall discover in our analysis, most doped semiconductor materials eventually become intrinsic at sufficiently high temperatures when the effect of impurities is swamped by other effects.

Figure 7.2 shows the difference between a noble metal, an insulator and both intrinsic and extrinsic semiconductors. The figure shows the donor states where the impurities donate electrons to the conduction band (*n*-type semiconductor) and the acceptor states which accept electrons from the valence band, thereby leaving holes in the valence band (*p*-type semiconductor). Clearly, the impurity states greatly influence the electrical and thermal properties since thermal excitation of electrons across the *full* width of the energy gap need not occur for conduction properties to be changed; all we need to do is supply sufficient thermal energy to raise electrons from the donor impurity atoms (ionize the donor atoms) into the conduction band—or raise electrons from the valence band into the acceptor states (leaving holes in the valence band) so that acceptor impurity atoms capture them. In either case the ionization energy is small and usually less than $1 \cdot 6 \times 10^{-20}$ J (less than 0·1 eV). We shall later show that because of this small ionization energy, substantial impurity atom ionization occurs at and around room temperature and it is just this impurity ionization which is the really important feature of the extrinsic semiconductor. The reader should understand that donor or acceptor impurity atoms lie in groups of very narrowly spaced energy states in the energy gap so that they may be treated as all possessing essentially the same energy. The reason for this is that

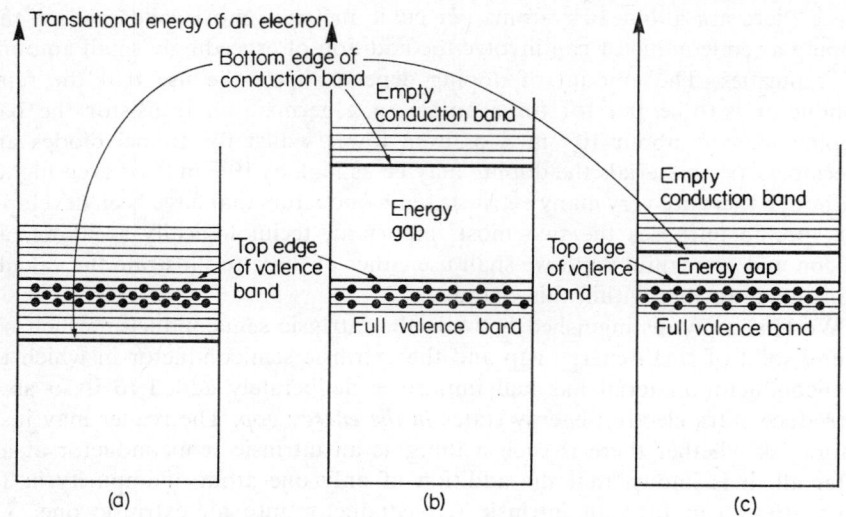

Figure 7.2. A schematic representation of some of the 'ladder' of allowed translational energy levels in:

Figure 7.2(a). A metal where the valence and conduction band energy levels overlap. The dots show the electron occupancy of the allowed levels at the absolute zero of temperature. The highest filled level is the Fermi level. This figure should be compared with Figure 4.2.

Figure 7.2(b). An insulator where there is a large energy gap separating the valence and conduction band. The dots show the electron occupancy of the allowed levels at the absolute zero. Note that the valence band is completely full (Pauli quota) of electrons whilst the conduction band is completely empty.

Figure 7.2(c). An intrinsic semiconductor where there is a small energy gap separating valence and conduction bands. The level occupancy at the absolute zero is as for Figure 7.2(b).

their concentration is so small compared with the concentration of the host atoms that the individual impurity atoms are separated by relatively large distances so that there is little coupling between the wave functions of 'neighbouring' impurity atoms and this in turn means that the donor and acceptor energy states are almost like those of isolated-atom energy state 'Bohr orbits' but with ionization energies characteristic of an atom in a dielectric rather than an atom in a vacuum. Some typical values of ionization energies are given in Table 7.1.

The Quasifree Thermally Perfect Electron Gas

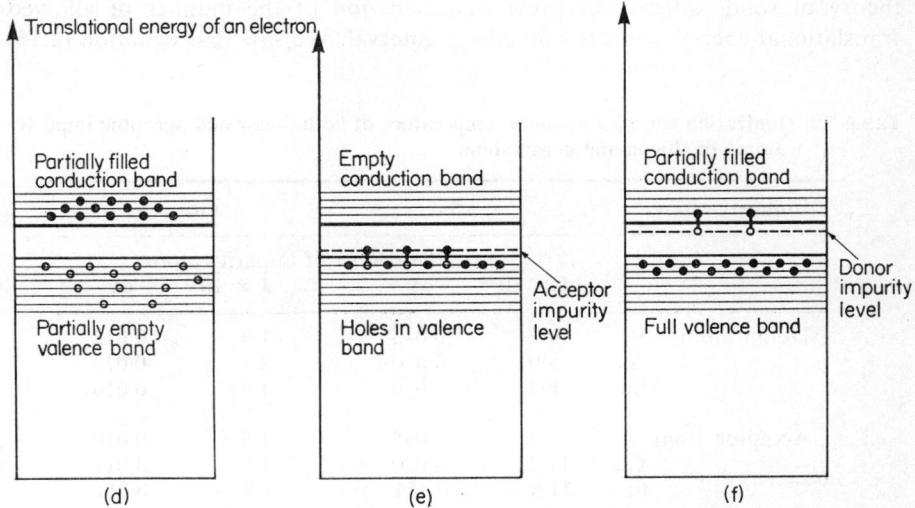

Figure 7.2(d). An intrinsic semiconductor at a finite non-zero temperature. Here some electrons have been thermally excited into the conduction band and have left an equal number of holes in the valence band. Both conduction band electrons and valence band holes are available for electrical and thermal conduction processes.

Figure 7.2(e). An extrinsic p-type semiconductor at a finite non-zero temperature. Here some electrons have been thermally excited into the acceptor impurity levels thereby leaving an equal number of holes in the valence band. Only these holes in the valence band are available for electrical and thermal conduction processes.

Figure 7.2(f). An extrinsic n-type semiconductor at a finite non-zero temperature. Here some donor impurity atoms have donated electrons by thermal excitation into the conduction band. Only the electrons in the conduction band are available for electrical and thermal conduction processes.

In general we see that the ionization energies of an n-type impurity in germanium is 0·01 eV (1·60 × 10^{21} J) and in silicon 0·05 eV (8·0 × 10^{-21} J) whilst for a p-type impurity the corresponding figures are 0·01 eV (1·60 × 10^{-21} J) and 0·06 eV (9·6 × 10^{-21} J).

7.2 The distribution of electron-energy states in a band

Let us now put the considerations of the previous section into mathematical form. Let us consider the valence band and determine how the allowed translational energy states are distributed as a function of their energy. The Bloch theory of solids affects our previous calculation of the number of allowed translational energy states in any energy interval. We saw that equation (2.14)

Table 7.1 Ionization energies at room temperature of both donor and acceptor impurity atoms in silicon and germanium.

Impurity atom		Si		Ge	
		\multicolumn{4}{c}{Ionization energy of impurity atom}
		J × 10²¹	eV	J × 10²¹	eV
Donor atom	P	7·2	0·045	1·9	0·012
	As	8·0	0·050	2·1	0·013
	Sb	6·4	0·040	1·6	0·010
Acceptor atom	B	7·2	0·045	1·6	0·010
	Ga	11·2	0·070	1·8	0·011
	In	24·6	0·154	1·8	0·011

gave us the density of states function which is the number of translational energy states in the energy range $d\varepsilon$ and if we recall that the Pauli principle allows up to two electrons in each of these translational energy states we obtain

Number of allowed single-particle translational states in the range ε to $\varepsilon + d\varepsilon$

$$= 2g(\varepsilon)d\varepsilon$$
$$= 4\pi(2m_e)^{\frac{3}{2}}Vh^{-3}\varepsilon^{\frac{1}{2}}d\varepsilon \tag{7.1}$$

We used this relation to derive equation (4.9) and plotted the result in Figure 4.4. However in band theory we must allow for the fact that only finite bands of translational energy levels are permitted, separated by an energy gap rather than a completely continuous array of translational energy levels. Let us put the translational energy at the bottom of the valence band as energy zero and that at the top of the valence band as energy ε_V. At the bottom of the valence band the number of allowed energy states follows equation (2.14), thus it varies as $\varepsilon^{\frac{1}{2}}$. At the top of the valence band the number of allowed translational energy states must fall to zero because we are approaching the energy gap, that is, the number of allowed translational energy states follows the form $(\varepsilon_V - \varepsilon)^{\frac{1}{2}}$. Figure 7.3 shows a schematic plot of the density of states function, the distribution of allowed translational energy states in the valence band. In the middle of the band there is a cusp due to crowding of the translational energy states; this cannot be explained simply. Of course this only considers

The Quasifree Thermally Perfect Electron Gas

the distribution of allowed translational energy states and if we require the electron numbers actually occupying these states then analogous to our derivation of equation (4.9), we must multiply the number of allowed single-particle translational energy states by the probability of actually finding an electron in any particular state. But what is this probability? Since we are dealing with electrons in a solid we must use Fermi–Dirac statistics regardless of whether the energy band is effectively unbounded as in Sommerfeld Theory or finite in

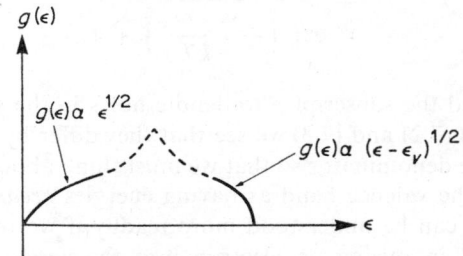

Figure 7.3. The density of states function $g(\varepsilon)$ which is the distribution of allowed carrier translational energy states in unit energy range of the valence band, as a function of the energy ε of the carrier.

width as in Bloch theory; electrons are fermions, and they obey the Pauli principle and Fermi–Dirac statistics.

Equation (4.2) showed that the F.D. distribution function for electrons is

$$\frac{N_n^*}{g_n} = \frac{1}{\exp\left(\dfrac{\varepsilon_n - \varepsilon_F}{kT}\right) + 1}$$

and in Section 4.4 we showed that the right-hand side of this equation is called the F.D. probability function $f(\varepsilon_n)$ since it gives the probability that any single-particle translational energy state of energy ε_n is actually occupied by an electron. Because of the essentially continuous distribution of translational energy states we re-wrote equation (4.8) as

$$\frac{N_n(\varepsilon)\mathrm{d}\varepsilon}{2g(\varepsilon)\mathrm{d}\varepsilon} = f_n(\varepsilon) = \frac{1}{\exp\left(\dfrac{\varepsilon - \varepsilon_F}{kT}\right) + 1} \tag{7.2}$$

where the subscript n on the left-hand side now refers to electrons in the conduction band and *not* to the translational quantum number.

At all temperatures above absolute zero there will be holes in the valence band due to some electrons having been thermally excited from the valence band into the conduction band. If $f_n(\varepsilon)$ is the probability that a single-particle translational state is actually occupied then $1 - f_n(\varepsilon)$ is the probability that there is *no* electron in that electronic state. But the absence of an electron is defined as the presence of a hole so that the probability distribution function for holes is

$$f_p(\varepsilon) = 1 - f_n(\varepsilon)$$
$$= \frac{1}{\exp\left(-\dfrac{\varepsilon - \varepsilon_F}{kT}\right) + 1} \quad (7.3)$$

where we have used the subscript p to denote holes in the valence band. If we compare equations (7.2) and (7.3) we see that they differ by a change in sign in the first term of the denominator so that we must think about holes lying deeply below the top of the valence band as having energies *greater* than those lying near the top. This can be understood more readily if we remember that more energy is involved in raising an electron into the conduction band from a deep-lying energy state in the valence band than from an energy state lying shallow; hence the hole so created in the valence band has more energy the deeper it lies. The fact that the kinetic energy of a hole increases the greater the 'descent' into the valence band gives rise to confusion in the literature regarding the choice of energy zero. It is quite common in textbooks to put energy zero at the top of the valence band rather than at the bottom as we have done; the writer believes the present description to be clearer.

7.3 Carrier concentrations in an intrinsic semiconductor in the classical approximation

Since any discussion of transport properties must involve, amongst other quantities, the carrier concentrations in the conduction and valence bands, let us calculate these concentrations for an intrinsic semiconductor. We have already shown that conduction occurs effectively either at the top of the valence band (holes) or the bottom of the conduction band (electrons); we therefore need only consider carrier concentrations in the vicinities of these band edges. Referring to Figure 7.4 let us consider the conduction band. In the previous section we showed that for the lower lying energy states close to the bottom of the conduction band, the density of states function is,

Number of allowed single-particle translational energy states in the energy range ε to $\varepsilon + d\varepsilon$ of the conduction band

$$= 2g(\varepsilon)d\varepsilon$$
$$= 4\pi(2m_n^*)^{\frac{3}{2}}Vh^{-3}(\varepsilon - \varepsilon_c)^{\frac{1}{2}}d\varepsilon$$

The Quasifree Thermally Perfect Electron Gas

Clearly when $\varepsilon = \varepsilon_c$, i.e. at the top of the energy gap, the equation above gives no allowed energy states.

Following equation (4.9) we have

Total number of electrons actually occupying single-particle translational energy states in the energy range ε to $\varepsilon + d\varepsilon$ of the conduction band

$$= N_n(\varepsilon)d\varepsilon$$
$$= 2g(\varepsilon)d\varepsilon \times f_n(\varepsilon)$$
$$= 4\pi(2m^*_{mn})^{\frac{3}{2}}Vh^{-3} \cdot \frac{(\varepsilon - \varepsilon_c)^{\frac{1}{2}} \cdot d\varepsilon}{\exp\left(\dfrac{\varepsilon - \varepsilon_F}{kT}\right) + 1} \quad (7.4)$$

Let us now make the 'classical' approximation, i.e. replace the F.D. distribution function by the M.B. distribution function. What is the justification for

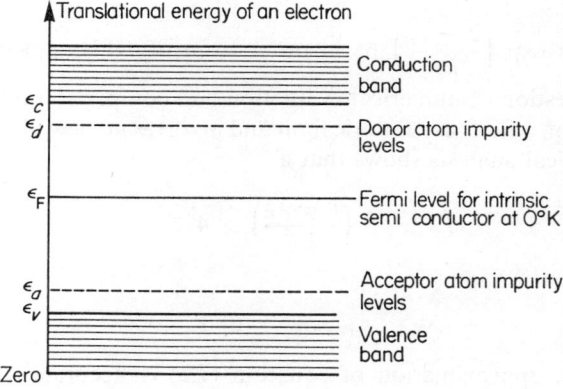

Figure 7.4. Schematic representation of some of the 'ladder' of allowed translational energy levels that make up both the valence and conduction bands of an extrinsic semiconductor. In addition the acceptor and donor impurity atom levels are also shown. No electrons or holes or ionized impurities are shown occupying the allowed energy levels although these will be present at all finite non-zero temperatures (see Figure 7.6).

doing this? In Section 4.3 we noted that in a *metal*, despite the fact that the bulk of the valence electrons obeyed F.D. statistics, the most important ones as regards electrical and thermal conduction processes were the conduction electrons, i.e. those in the high-energy tail of the distribution and that to all intents and purposes these conduction electrons in the high-energy tail obeyed M.B. statistics. The reader should note that there is no contradiction here.

The high-energy tail electrons actually do obey F.D. statistics but as we have noted in Section 2.13 there is essentially no difference in the high-energy tails of the F.D. and M.B. distribution so that we can regard the high-energy tail as *effectively* M.B. Now equation (7.2) for the conduction-band electron distribution function is the F.D. distribution,

$$f_n(\varepsilon) = \frac{1}{\exp\left(\dfrac{\varepsilon - \varepsilon_F}{kT}\right) + 1}$$

In Section 2.10 we showed that when $\exp\left(\dfrac{\varepsilon - \varepsilon_F}{kT}\right) \gg 1$ then we obtained the M.B. distribution [see equation (2.40)]. Hence the distribution function becomes,

$$f_n(\varepsilon) \sim \exp\left(-\frac{\varepsilon - \varepsilon_F}{kT}\right) \tag{7.5}$$

How big must $\exp\left(\dfrac{\varepsilon - \varepsilon_F}{kT}\right)$ be in order to justify this approximation? This is purely a question of numerical evaluation and comparison between a mathematical function and its asymptotic limit and involves no new physical reasoning. Such a numerical analysis shows that if

$$\left(\frac{\varepsilon - \varepsilon_F}{kT}\right) > 4$$

i.e.

$$\varepsilon - \varepsilon_F > 4kT \tag{7.6}$$

then the M.B. approximation of equation (7.5) is 'accurate'. But what is the meaning of equation (7.6)? Consider, for example, the conduction band with its lower lying translational energy states occupied by electrons. The lowest value of kinetic energy for electrons in the conduction band is ε_c; hence equation (7.6) means that if the conduction-band electrons are to obey M.B. statistics then we must have

$$\varepsilon_c - \varepsilon_F > 4kT$$

i.e.

$$\varepsilon_c > \varepsilon_F + 4kT \tag{7.7}$$

which tells us that if the Fermi level lies further away, lower in energy than $4kT$, from the bottom of the conduction band, then the electron distribution is M.B. Clearly in this case the electrons are not degenerate (as they certainly *are* for the bulk of the valence electrons in a metal) and they obey M.B. statistics so that we have a non-degenerate intrinsic semiconductor. The reader may ask what happens if equation (7.7) is not obeyed. The answer is that if the Fermi level lies *within* $4kT$ of the bottom of the conduction band edge then the semiconductor is said to be partially degenerate whilst if the Fermi level rises even higher so that it actually enters the conduction band then the semiconductor is said to be degenerate and the electrons obey full F.D. statistics. Clearly this

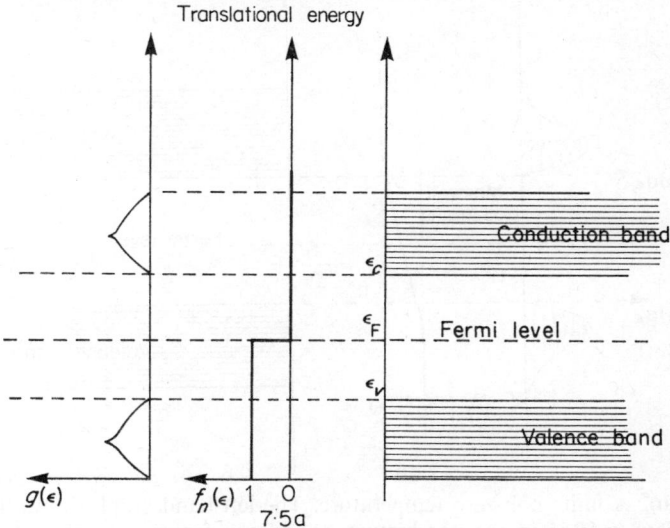

Figure 7.5. The valence and conduction band 'ladder' of translational energy levels of an intrinsic semiconductor at:

Figure 7.5(a). Absolute zero of temperature. The left-hand graph of the figure shows the number of allowed translational energy states $g(\varepsilon)$ in unit energy range ($\mathrm{d}\varepsilon = 1$) for both *p*-type carriers in the valence band and *n*-type carriers in the conduction band (this graph is identical to Figure 7.3 rotated anticlockwise through 90°). The right-hand graph of the figure shows the F.D. probability distribution function for electrons $f_n(\varepsilon)$ which will be seen to have a value of zero throughout the entire conduction band since this band is empty (this graph is identical to the $T = 0°\mathrm{K}$ graph of Figure 4.3 rotated anticlockwise through 90°. The Fermi level is in the middle of the energy gap. No electrons are shown in the valence band which is in fact full and there are no holes in this band.

raises the question as to where the Fermi level in an intrinsic semiconductor is and this very important question we shall answer shortly.

In the meantime let us assume that the classical approximation is valid— an assumption that is correct in very many semiconductor applications. Equation (7.4) becomes

$$N_n(\varepsilon)\mathrm{d}\varepsilon = 4\pi(2m_n^*)^{\frac{3}{2}}Vh^{-3}(\varepsilon - \varepsilon_c)^{\frac{1}{2}} \exp\left(-\frac{\varepsilon - \varepsilon_F}{kT}\right) \mathrm{d}\varepsilon$$

Figure 7.5 shows a plot of this distribution of electrons over the allowed translational energy states in the conduction band. We note from this figure that the electrons contributing significantly to conduction processes have energies near the bottom of the conduction band, as we have already remarked.

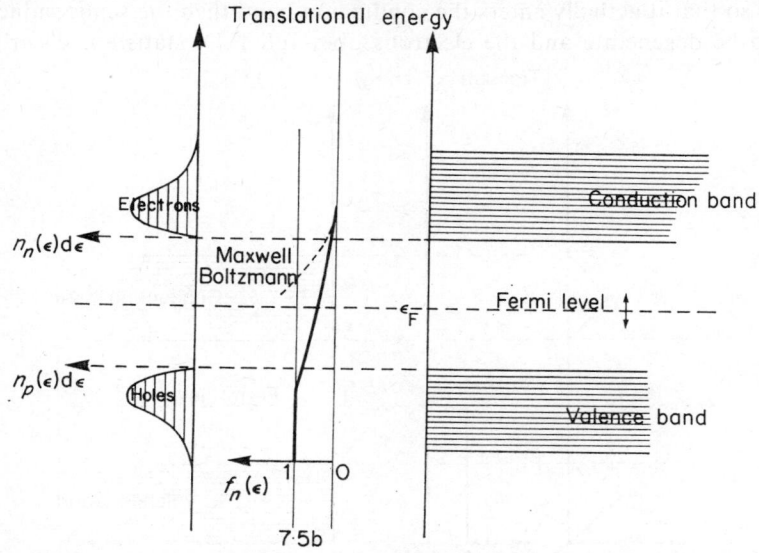

Figure 7.5(b). A finite non-zero temperature. The left-hand graph shows the electron concentrations $n_n(\varepsilon)d\varepsilon$ in the conduction band as a function of their translational kinetic energy and also the hole concentrations $n_p(\varepsilon)d\varepsilon$ in the valence band, likewise as a function of their kinetic energy. These two graphs should be compared with the $T > 0°K$ graph of Figure 4.4 rotated anticlockwise through 90°. The right-hand graph shows the F.D. electron probability distribution function $f_n(\varepsilon)$ which is now non-zero in the conduction band. (This graph is identical to the finite non-zero temperature graph of Figure 4.3 rotated anticlockwise through 90°. The Fermi level is marked with a two-way arrow since this can move about the energy gap depending on the temperature and the effective mass ratio of electrons to holes. No electrons or holes are shown occupying the allowed energy levels although these are present in the manner shown in Figure 7.2(d).

A mirror-image distribution holds for holes in the valence band; this must be so in order to preserve charge neutrality (to which we shall return later).

In the classical approximation equation (7.4) becomes on integration

Total number of electrons actually occupying single-particle translational energy states in the range ε_c to the upper limit of energy of the conduction band

$$= 4\pi(m_n^*)^{\frac{3}{2}} V h^{-3} \int_{\varepsilon}^{\text{Upper Limit}} (\varepsilon - \varepsilon_c)^{\frac{1}{2}} \exp\left(-\frac{\varepsilon - \varepsilon_F}{kT}\right) d\varepsilon$$

The Quasifree Thermally Perfect Electron Gas

i.e.

$$N_n = 4\pi(2m_n^*)^{\frac{3}{2}} V h^{-3} \exp\left(\frac{\varepsilon_F}{kT}\right) \cdot \int_{\varepsilon_c}^{\text{Upper Limit}} (\varepsilon - \varepsilon_c)^{\frac{1}{2}} \exp\left(-\frac{\varepsilon}{kT}\right) d\varepsilon \quad (7.8)$$

In the above equations we have replaced the symbol m_e^* by m_n^* to show that we are dealing with *n*-type charge carriers (electrons). Likewise we shall denote the total number of electrons in the conduction band by N_n rather than by N_e.

The integral in equation (7.8) rapidly converges due to the exponential function. This rapid convergence is a consequence of the fact that the probability of actually finding an electron in any allowed translational energy state in the conduction band decreases very rapidly (exponentially) the higher the energy of that state. This means that the upper translational energy states of the conduction band are effectively empty so that we can take the upper limit of the integral as infinity rather than the energy appropriate to the top of the conduction band; thus equation (7.8) becomes

$$N_n \sim 4\pi(2m_n^*)^{\frac{3}{2}} V h^{-3} \exp\left(\frac{\varepsilon_F}{kT}\right) \cdot \int_{\varepsilon_c}^{\infty} (\varepsilon - \varepsilon_c)^{\frac{1}{2}} \exp\left(-\frac{\varepsilon}{kT}\right) d\varepsilon \quad (7.9)$$

Within the integral let us make the substitution $x = \varepsilon - \varepsilon_c$; then equation (7.9) becomes

$$N_n = 4\pi(2m_n^*)^{\frac{3}{2}} \exp\left(\frac{\varepsilon_F}{kT}\right) V h^{-3} \int_0^{\infty} x^{\frac{1}{2}} \exp\left(-\frac{x + \varepsilon_c}{kT}\right) dx$$

$$= 4\pi(2m_n^*)^{\frac{3}{2}} \exp\left(\frac{\varepsilon_F - \varepsilon_c}{kT}\right) V h^{-3} \int_0^{\infty} x^{\frac{1}{2}} \exp\left(-\frac{x}{kT}\right) dx$$

But the definite integral has the value $\dfrac{\sqrt{\pi}}{2}(kT)^{\frac{3}{2}}$

∴

$$N_n = \frac{2(2\pi m_n^* kT)^{\frac{3}{2}} V}{h^3} \exp\left(-\frac{\varepsilon_c - \varepsilon_F}{kT}\right) \quad (7.10)$$

This gives the total number of electrons in the conduction band.

This result, if viewed in the light of the formulae derived in Chapters 2 and 3, has a familiar look. Equation (2.58) shows that the partition function for the F.D. distribution expressed in terms of M.B. distribution is

$$Z_{FD} = \exp \alpha \cdot Z_{MB}$$

$$= \exp\left(\frac{\varepsilon_F}{kT}\right) \cdot Z_{MB}$$

using equation (4.1) that relates the undetermined multiplier α to the Fermi energy. However, from equation (2.58) we have

$$Z_{\text{MB}} = \text{`} \Sigma g_n \exp\left(-\frac{\varepsilon_n}{kT}\right) \text{'}$$

We have put the right-hand side in inverted commas because we must be careful when dealing with the conduction band, to remember that the allowed energy states do not start from $\varepsilon_n = 0$ but in fact from $\varepsilon_n = \varepsilon_c$ up to higher energies, i.e.

$$Z_{\text{MB}} = \Sigma g_n \exp\left(-\frac{\varepsilon_c + \varepsilon_n}{kT}\right) \tag{7.11}$$

is the correct expression for the M.B. partition function for the electrons in the conduction band. Clearly we can write this as

$$Z_{\text{MB}} = \exp\left(-\frac{\varepsilon_c}{kT}\right) \Sigma g_n \exp\left(-\frac{\varepsilon_n}{kT}\right) \tag{7.12}$$

But from equation (3.14) the electronic partition function for M.B. electrons treated as a thermally perfect gas has a value

$$\Sigma g_n \exp\left(-\frac{\varepsilon_n}{kT}\right) = \frac{2(2\pi m_n^* kT)^{\frac{3}{2}} V}{h^3}$$

We can therefore combine equations (7.11) (7.12) and (3.14) to give

$$Z_{\text{FD}} = \exp\left(\frac{\varepsilon_F}{kT}\right) \cdot \exp\left(-\frac{\varepsilon_c}{kT}\right) \cdot \frac{2(2\pi m_n^* kT)^{\frac{3}{2}} V}{h^3}$$

$$= \frac{2(2\pi m_n^* kT)^{\frac{3}{2}} V}{h^3} \exp\left(-\frac{\varepsilon_c - \varepsilon_F}{kT}\right) \tag{7.13}$$

However, from equation (2.53) we have

$$N_n = \left(\frac{\partial Z_{\text{FD}}}{\partial \alpha}\right)_{T,V}$$

$$= \left[\frac{\partial Z_{\text{FD}}}{\partial \left(\frac{\varepsilon_F}{kT}\right)}\right]_{T,V}$$

$$= kT \left(\frac{\partial Z_{\text{FD}}}{\partial \varepsilon_F}\right)_{T,V}$$

$$= kT \frac{\partial}{\partial \varepsilon_F} \left[\frac{2(2\pi m_n^* kT)^{\frac{3}{2}} V}{h^3} \exp\left(-\frac{\varepsilon_c - \varepsilon_F}{kT}\right)\right]$$

$$= 2(2\pi m_n^* kT)^{\frac{3}{2}} V h^{-3} \exp\left(-\frac{\varepsilon_c - \varepsilon_F}{kT}\right)$$

The Quasifree Thermally Perfect Electron Gas

which is equation (7.10). We have therefore shown that our present 'classical' analysis agrees (as it should!) with our earlier discussion of the limiting form of F.D. distribution, the M.B. distribution.

Equation (7.10) at first sight appears to give us all the information that we require to determine the numerical value of the concentration of electrons in the conduction band in terms of the temperature of the semiconductor, the energy of the bottom of the conduction band edge ε_c and the Fermi energy ε_F. Equation (7.10) is often written in terms of the electron concentration (number density) n_n

$$n_n = \frac{N_n}{V}$$

$$= 2(2\pi m_n^* kT)^{\frac{3}{2}} h^{-3} \exp\left(-\frac{\varepsilon_c - \varepsilon_F}{kT}\right)$$

$$= n_c \exp\left(-\frac{\varepsilon_c - \varepsilon_F}{kT}\right) \quad (7.14)$$

Here

$$n_c = 2(2\pi m_n^* kT)^{\frac{3}{2}} h^{-3} \quad (7.15)$$

$$= 4\cdot 81 \times 10^{21} T^{\frac{3}{2}} \text{ per cubic metre}$$

$$= 2\cdot 50 \times 10^{25} \text{ per cubic metre at } T = 300°K \quad (7.16)$$

if we take the effective electron mass in the conduction band equal to that of an isolated free electron. This is, in fact, usually *not* a good approximation because as we have already remarked the electrons that we are concerned with in the conduction band lie fairly close to the bottom edge of the conduction band because of the 'small' number of valence-band electrons actually thermally excited into the conduction band. Although the conduction-band electrons are essentially free, because they lie in allowed energy states close to the bottom edge of the band, quantum theory states that they will have small effective masses, i.e. $\frac{m_n^*}{m_e} \ll 1$. It should be noted that the multiplier 2 that is the first term on the right-hand side of equation (7.15) only occurs when every translational energy state is occupied by two electrons of opposite spin; we return to this point below.

By analogous reasoning we can readily find an expression for the hole concentration in the valence band, viz.

$$n_p = \frac{N_p}{V}$$

$$= 2(2\pi m_p^* kT)^{\frac{3}{2}} h^{-3} \exp\left(-\frac{\varepsilon_F - \varepsilon_V}{kT}\right) \tag{7.17}$$

$$= n_V \exp\left(-\frac{\varepsilon_F - \varepsilon_V}{kT}\right) \tag{7.18}$$

where m_p^* is the effective mass of a hole in the valence band.

Equations (7.14) and (7.17) are the important equations that show the very great difference between metals and semiconductors. We have seen from Chapter 4 that the number of valence electrons per unit volume *is a constant*, independent of temperature in the case of a metal (in passing we should note that this did *not* mean that the electrical conductivity or thermal conductivity of a metal was likewise independent of temperature because, as we have already seen, these transport coefficients depend not only on the valence-electron concentration but also on the scattering processes that the conduction electrons suffer and the number of scattering processes suffered by a conduction electron is temperature-dependent). However, in the present case of the semiconductor we see that the number of 'free' charge carriers is anything but constant; rather it increases exponentially, i.e. exceedingly *rapidly* with increasing temperature.

7.4 The Law of Mass Action and the Fermi level in an intrinsic semiconductor

We have now obtained expressions for the concentrations of charge carriers in a band as a function of temperature, the respective energies of the bottom edge of the conduction band or the top edge of the valence band—and the Fermi energy. If we multiply equations (7.14) and (7.18) together we get

$$n_n n_p = 4(2\pi kT)^3 h^{-6} (m_n^* m_p^*)^{\frac{3}{2}} \exp\left(-\frac{\varepsilon_c - \varepsilon_v}{kT}\right) \tag{7.19}$$

However, from Figure 7.4 it is obvious that

$$\varepsilon_c - \varepsilon_V = \varepsilon_G \tag{7.20}$$

where ε_G is the width of the energy gap between the bottom of the conduction band and the top of the valence band. In addition, in an intrinsic semiconductor, the number of holes in the valence band must equal the number of electrons in the conduction band since holes can only be created in the valence band by electrons being raised into the conduction band (this statement only holds for an intrinsic semiconductor and a very different result holds for an extrinsic semiconductor, as we shall shortly see), i.e. in an intrinsic semiconductor,

$$n_n = n_p \tag{7.21}$$

The Quasifree Thermally Perfect Electron Gas

Let us put

$$n_n = n_p = n_i \qquad (7.22)$$

where n_i = concentration of carriers of either sign in their appropriate band in an intrinsic semiconductor, i.e. n_i is the intrinsic carrier concentration. Hence equation (7.19) becomes

$$n_i^2 = n_n n_p = 4(2\pi kT)^3 h^{-6}(m_n^* m_p^*)^{\frac{3}{2}} \exp\left(-\frac{\varepsilon_G}{kT}\right)$$

i.e.

$$n_i = n_n = n_p = \sqrt{n_n n_p} = 2(2\pi kT)^{\frac{3}{2}} h^{-3}(m_n^* m_p^*)^{\frac{3}{4}} \exp\left(-\frac{\varepsilon_G}{2kT}\right) \qquad (7.23)$$

This is the Law of Mass Action. This result is in some ways analogous to the Saha equation [see equation (3.26)], since the Saha equation gives the concentration of electrons and ions in a gaseous plasma as a function of the pressure, temperature and ionization potential of the atoms whilst equation (7.23) gives the concentration of electrons and holes in the conduction and valence bands of an intrinsic semiconductor as a function of the effective masses, temperature and energy gap of the semiconductor material (clearly the energy gap is equivalent to an ionization potential since it represents the energy that a valence-band electron must gain from the lattice thermal energy to be freed completely from the valence band, so that it can cross the energy gap and enter the conduction band).

The great usefulness of equation (7.23) is that it does not contain the Fermi energy, that is, we can calculate the intrinsic carrier concentrations without a knowledge of the Fermi energy.

This is different from the case of a metal. In Chapter 4 we gave a formula for determining the Fermi energy in a metal, viz. equation (4.7b) as

$$\varepsilon_F = \frac{h^2}{8m_e}\left(\frac{3}{\pi} \cdot n_e\right)^{\frac{2}{3}}$$

However, this defines the Fermi energy in terms of the valence-electron concentration in the metal (and we note that it is independent of temperature). Of course we could rearrange this formula to read, as we did in equation (4.10)

$$n_e = \frac{N_e}{V} = \frac{8\pi}{3h^3}(2m_e)^{\frac{3}{2}} \varepsilon_F^{\frac{3}{2}}$$

which gives the valence-electron concentration in terms of Fermi energy. We therefore conclude that we can either deduce the Fermi energy from a known value of valence-electron concentration or deduce the valence-electron concentration from a known Fermi energy but we cannot deduce both quantities together without a knowledge of either one or the other.

We can further simplify equation (7.23) by writing

$$m_n^* \sim m_p^*$$

where

$$m_n^* \ll m_e$$

This is a good approximation since the electrons and holes that participate in the conduction process lie close to their respective band edges and their effective masses are equal. In this case our intrinsic semiconductor equations become

$$\left.\begin{array}{l} n_n = n_p = n_i = n_c \exp\left(-\dfrac{\varepsilon_G}{2kT}\right) \\[6pt] = n_V \exp\left(-\dfrac{\varepsilon_G}{2kT}\right) \end{array}\right\} \quad (7.23a)$$

Despite the fact that we do not need to know the position of the Fermi level in order to calculate the intrinsic carrier concentrations, we shall find in our discussion of the extrinsic semiconductor that we do, in fact need to know its position in the intrinsic semiconductor. This presents no problem because we already have sufficient formulae available to determine its position. Before doing so let us see if we can deduce where it should lie based on our earlier discussion of the F.D. distribution function. Equation (7.2) is

$$f_n(\varepsilon) = \dfrac{1}{\exp\left(\dfrac{\varepsilon - \varepsilon_F}{kT}\right) + 1}$$

and we know from Chapter 4 that in a metal the Fermi energy is that energy that makes $f(\varepsilon) = 0.5$—see equation (4.4). This followed because at the absolute zero for a metal

$$f(\varepsilon) = 1 \text{ for all states of energy } \varepsilon \leqslant \varepsilon_F$$
$$f(\varepsilon) = 0 \text{ for all states of energy } \varepsilon > \varepsilon_F$$

Strictly speaking, at the absolute zero the Fermi level in a metal should be that energy state that is *half way* between the highest filled (uppermost occupied) translational energy level and the lowest empty translational energy level immediately above it in order to get $f(\varepsilon) = 0.5$. However, since the translational levels energy are essentially continuously distributed we can take the Fermi energy as the energy of the highest filled translational energy level without loss of accuracy.

However, in a semiconductor the allowed translational energy levels are not continuous other than *within* the valence band and *within* the conduction

The Quasifree Thermally Perfect Electron Gas

band because these bands are separated by the energy gap. Furthermore, at the absolute zero the valence band is completely full so that $f_n(\varepsilon) = 1$ for all valence electrons of energy $\varepsilon \leqslant \varepsilon_V$; also the conduction band is completely empty, so that $f_n(\varepsilon) = 0$ for all energies $\varepsilon \geqslant \varepsilon_c$. Therefore there are no occupied energy levels for which $f_n(\varepsilon) = 0.5$. This must mean that the Fermi level cannot coincide with an allowed translational energy level so that it must lie within the forbidden energy states where this forbidden state must have an energy between ε_V and ε_c in order to give $f_n(\varepsilon) = 0.5$, i.e. the Fermi level must lie in the energy gap. But where about in the energy gap? We can answer this using the fact that $f_n(\varepsilon) = 0.5$ gives the Fermi energy at the absolute zero. What we do is to determine the energy of the highest occupied or 'filled' translational energy level $\varepsilon_{\text{filled}}$ and also determine the energy of the lowest unoccupied or 'empty' translational level *above* $\varepsilon_{\text{filled}}$ which has energy $\varepsilon_{\text{empty}}$. Then at the absolute zero

$$\varepsilon_F = \frac{\varepsilon_{\text{empty}} + \varepsilon_{\text{filled}}}{2} \tag{7.24}$$

For the intrinsic semiconductor at the absolute zero we have $\varepsilon_{\text{empty}} \equiv \varepsilon_c$ and $\varepsilon_{\text{filled}} \equiv \varepsilon_V$
whence

$$\varepsilon_F = \frac{\varepsilon_c + \varepsilon_V}{2} \tag{7.25}$$

i.e. at the absolute zero the Fermi level lies exactly in the middle of the energy gap. What happens if the temperature of the intrinsic semiconductor rises above absolute zero? In this case equation (7.25) does not hold and we have to turn to the equations already derived, equations (7.14) and (7.18) which, when divided, yield

$$\frac{n_n}{n_p} = \frac{2(2\pi m_n^* kT)^{\frac{3}{2}} V h^3}{2(2\pi m_p^* kT)^{\frac{3}{2}} V h^3} \cdot \frac{\exp\left(-\dfrac{\varepsilon_c - \varepsilon_F}{kT}\right)}{\exp\left(-\dfrac{\varepsilon_F - \varepsilon_V}{kT}\right)}$$

$$\therefore \quad \frac{n_n}{n_p} = \left(\frac{m_n^*}{m_p^*}\right)^{\frac{3}{2}} \exp\left\{\frac{2\varepsilon_F - (\varepsilon_c + \varepsilon_V)}{kT}\right\} \tag{7.26}$$

But from equation (7.22) the left-hand side of equation (7.26) has the value unity whence the Fermi energy is

$$\varepsilon_F = \frac{\varepsilon_c + \varepsilon_V}{2} + \frac{3}{4} kT \ln\left(\frac{m_p^*}{m_n^*}\right) \tag{7.27}$$

At the absolute zero, equation (7.27) gives equation (7.25). In fact equation (7.27) shows that equation (7.25) holds accurately for all temperatures (not just the absolute zero) for an intrinsic semiconductor provided that

$$m_n^* \sim m_p^* \tag{7.28}$$

where

$$m_n^* \ll m_e \tag{7.29}$$

In addition, because the second term on the right-hand side of equation (7.27) is a logarithmic one, slight differences in the effective masses do not affect this term to any extent anyway.

Figure 7.5 summarizes the intrinsic semiconductor both at the absolute zero and at some finite non-zero temperature. We see clearly in Figure 7.5(b) that at this non-zero temperature the 'tail' of the F.D. distribution reaches into the conduction band. This must be so because some electrons will have gained sufficient thermal energy from the lattice in order to be thermally excited from the valence band into the conduction band. Any thermal energy that these electrons gain from the lattice that is in excess of ε_c will be kinetic energy in the conduction band. Figure 7.5(b) also shows the M.B. distribution and we note that the F.D. and M.B. high-energy 'tails' merge, thereby confirming the classical approximation.

Let us use equation (7.23a) to calculate the concentration of electrons in the conduction band of a typical intrinsic semiconductor. Table 7.2 gives values of the energy gap for several semiconductors. There appears to be a considerable variation in the values of the energy gap quoted in the literature (as much as 100% in some cases!) so that the values tabulated in Table 7.2 should not be treated as exact. In addition the energy gap is a function of temperature and very often values given in the literature omit mention of the temperature altogether.

Table 7.2 Values of the energy gap at 300°K and melting points of common intrinsic semiconductors.

Semiconductor	Energy gap at 300°K (approximate)		Melting point °K
	J	eV	
Si	1.84×10^{-19}	1·15	1690
Ge	1.04×10^{-19}	0·65	1235
GaSb	1.25×10^{-19}	0·78	975
InSb	0.37×10^{-19}	0·23	796
GaAs	2.24×10^{-19}	1·4	1553
PbTe	0.48×10^{-19}	0·30	1190

The Quasifree Thermally Perfect Electron Gas

Consider the case of germanium at 300°K

$$n_i = n_c \exp\left(-\frac{\varepsilon_G}{2kT}\right)$$

$$= 2{\cdot}50 \times 10^{25} \exp\left(-\frac{1{\cdot}04 \times 10^{-19}}{2 \times 1{\cdot}38 \times 10^{-23} \times 300}\right) m^{-3}$$

$$= 8{\cdot}77 \times 10^{19} \text{ electrons per cubic metre} \tag{7.30}$$

But there are about $4{\cdot}5 \times 10^{28}$ atoms per cubic metre in germanium; hence in a typical intrinsic semiconductor there is only one conduction electron in the conduction band (and therefore one hole in the valence band) for every 10^8 germanium atoms. This calculation also tells us something about the level of doping required to make an intrinsic semiconductor into an extrinsic one. If we add about one atom of donor impurity to every 10^7 germanium atoms we clearly will have a profound effect on the number of electrons in the conduction band provided these donor atoms are ionized. If we compare equation (7.30) with the case of a metal then in chapter 4 we found that for monovalent metals there were about 5×10^{28} electrons per cubic metre. Hence the prefix 'semi' in semiconductor is well chosen. If we perform a calculation similar to equation (7.30) for the case of silicon at room temperature, we find that the intrinsic electron concentration is $6{\cdot}25 \times 10^{15}$ per cubic metre. These *calculated* values of intrinsic electron concentration do not agree with measured values, because they have both been based on using the mass of an isolated electron for the effective mass of an electron in an energy band. (Also any calculated values of intrinsic carrier concentration will be very sensitive to the chosen value of the energy gap which, as we have already remarked, is temperature-dependent.) *Experiment* shows that the intrinsic carrier concentration at room temperature in silicon is about $1{\cdot}4 \times 10^{16}$ per cubic metre and that for germanium $2{\cdot}5 \times 10^{19}$ per cubic metre. We also note that if we wish to observe true intrinsicity in semiconductors at room temperature then we must ensure that there are no impurities which can give rise to ionized impurity concentrations greater than about 10^{18} per cubic metre. We see that for silicon at room temperature $n_i < n_d^+$ so that silicon is *not* intrinsic whereas in the case of germanium we have $n_i > n_d^+$ so that germanium can be intrinsic at room temperature where n_d^+ is the concentration of ionized donor atoms (see Section 7.5).

It must be remembered that the intrinsic carrier concentration is critically dependent (exponentially) on temperature. Germanium can be operated up to 1000°K and above (see Table 7.2). At 1000°K we can readily calculate, in exactly the same manner as above, that the conduction-band electron concentration becomes $5{\cdot}60 \times 10^{23}$ per cubic metre, thus there is a substantial increase in the conduction-band electron concentration of about 10^4 for a temperature rise of 700 deg C. This high-temperature calculation presupposes that the energy gap is temperature-independent, which however is not the case. In fact, this calculation is likely unrealistic altogether because the temperature of the semiconductor must always remain below that level at which the solders used in semiconductor fabrication soften. The maximum operating temperature for germanium transistors is about 360°K whilst that for silicon about 480°K (see Section 7.6).

Equation (7.23a) can also be used to confirm that M.B. statistics is a good approximation for intrinsic semiconductor charge carriers. In Section 2.13 we introduced the degeneracy parameter

$$\Lambda = \frac{(2\pi mkT)^{\frac{3}{2}} V}{h^3 \cdot N}$$

and showed in equation (2.75) that provided $\Lambda \gg 1$ then M.B. statistics were applicable (where once again we stress that M.B. statistics is not a new type of statistics; it is merely the limiting form of F.D. statistics that occurs when there are very many more allowed carrier translational energy states available than there are carriers to occupy them).

Equation (7.23a) can be written

$$n_i = \frac{N_i}{V} = \frac{2(2\pi m^* kT)^{\frac{3}{2}}}{h^3} \exp\left(-\frac{\varepsilon_G}{2kT}\right)$$

i.e.

$$\Lambda = \frac{(2\pi m^* kT)^{\frac{3}{2}}}{h^3} \cdot \frac{V}{N_i} = \frac{1}{2} \exp\left(\frac{\varepsilon_G}{2kT}\right)$$

Using the data given in Table 7.2 we find that at room temperature

$$\Lambda_{Si} = 1 \cdot 7 \times 10^9$$
$$\Lambda_{Ge} = 8 \times 10^4$$

so that in both cases $\Lambda \gg 1$, thereby further confirming that our formulae are consistent with the classical approximation.

7.5 Carrier concentrations in an extrinsic semiconductor

Our derivation of the Law of Mass Action equation (7.23) nowhere assumed that the conductor was necessarily intrinsic; it merely concerned itself with the concentration of charge carriers n_n and n_p in the conduction and valence bands and did not 'enquire' how these carriers got into these bands. We know that in the case of an intrinsic semiconductor, electrons can only reach the conduction band by being thermally excited from the valence band, that is, they pick up sufficient thermal energy from the lattice to cross the energy gap and enter the conduction band. However, in an extrinsic semiconductor,

The Quasifree Thermally Perfect Electron Gas

electrons can also enter the conduction band from the donor-atom impurity levels.

Equation (7.23) showed that for an intrinsic semiconductor

$$n_i = n_n = n_p = \sqrt{n_n n_p}$$

because electrons and holes arise in pairs. However, this is *not* the case in an extrinsic semiconductor, where

$$n_n \neq n_p \tag{7.31}$$

In fact the whole technological importance of the extrinsic semiconductor is that we can arrange for a predominance of electron carriers in the conduction band *or* hole carriers in the valence band according to the dopant used. Of course in any practical extrinsic semicondutor, because of the finite limitations of refining techniques, both donor and acceptor impurity atoms will likely be present but we can always ensure that one impurity type dominates the other.

In an analysis of the extrinsic semiconductor we must consider:

a. Electrons in the conduction band that have come from the valence band; these are electrons that would be in the conduction band even if all impurities were absent in the semiconductor and they are there as a result of thermal excitation across the full width of the energy gap.
b. Electrons in the conduction band that have come from donor impurity atoms which have ionized, that is, lost an electron to the conduction band.

The total concentration of electrons in the conduction band is the sum of (a) and (b) and is n_n.
c. Holes in the valence band that occur because electrons have left the valence band due to thermal excitation, crossed the energy gap and entered the conduction band. These holes would be present if all impurities were absent from the semiconductor.
d. Holes in the valence band that occur because acceptor impurity atoms have accepted electrons from the valence band and consequently have left holes in the valence band.

The total concentration of holes in the valence band is the sum of (c) and (d) and is n_p.

We therefore have the situation depicted in Figure 7.6. There are two restrictions on n_n and n_p. These are:

Restriction 1. The Law of Mass Action holds

$$n_n = n_c \exp\left(-\frac{\varepsilon_c - \varepsilon_F}{kT}\right) \tag{7.32}$$

$$n_p = n_V \exp\left(-\frac{\varepsilon_F - \varepsilon_v}{kT}\right) \tag{7.33}$$

However,
$$n_n \neq n_p$$
but,
$$n_n n_p = \sqrt{n_c n_v} \cdot \exp\left(-\frac{\varepsilon_G}{2kT}\right) \tag{7.34}$$
$$= n_i^2 \tag{7.35}$$

The reader must appreciate that equations (7.32) and (7.33) are identical with equations (7.14) and (7.18). Where then does the extrinsic nature of the semiconductor enter the argument? The answer must be that the *Fermi energy* ε_F *is*

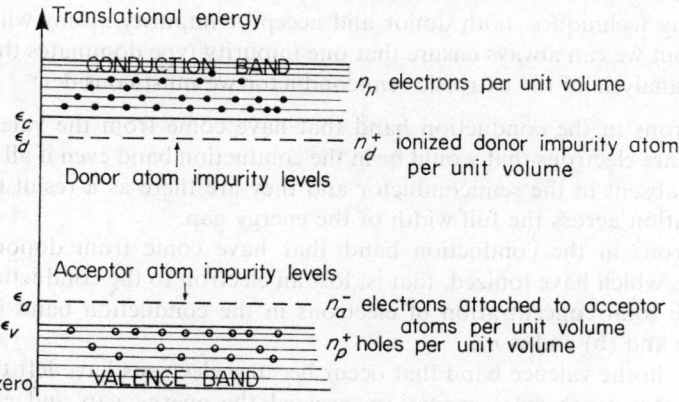

Figure 7.6. Schematic representation of the population of charge carriers and immobile ionized impurity atoms in the translational energy levels shown in Figure 7.4 for an extrinsic semiconductor at a finite non-zero temperature. In general the electron concentration in the conduction band is not equal to the hole concentration in the valence band. These charge carriers in their respective bands are present because of the types of thermal excitation processes shown in Figure 7.2(e) and 7.2(f).

dependent on the level of doping; another way in which extrinsicity arises is shown in equation (7.31). Equation (7.35) dramatically shows the effect of doping. We have seen in the previous section that for germanium at room temperature the calculated intrinsic carrier concentration is $8 \cdot 8 \times 10^{19}$ electrons per cubic metre, say 10^{20} m^{-3}. Suppose we add donor-impurity atoms so that the concentration of donor atoms which are ionized is $n_d^+ = 10^{23}$ per cubic metre. Then
$$n_n = n_i + n_d^+$$
$$= 10^{20} + 10^{23}$$
$$= 10^{23} \text{ m}^{-3}$$

But from equation (7.35)

$$n_p = \frac{n_i^2}{n_n}$$
$$= \frac{10^{40}}{10^{23}}$$
$$= 10^{17} \text{ m}^{-3}$$

Hence by the addition of an extremely small amount of donor-impurity atoms (1 in 10^5 host atoms) we have raised the electron concentration in the conduction band from an intrinsic value of about 10^{20} electrons per cubic metre to 10^{23} electrons per cubic metre and reduced the hole concentration in the valence band to 10^{17} holes per cubic metre.

It is important to stress that impurities that do not ionize, that is, donor atoms that do not give an electron to the conduction band and acceptor atoms that do not accept an electron from the valence band *have no effect on the carrier concentration*. It is not sufficient, therefore, for us to merely quote the doping level as 'so many impurity atoms per cubic metre'; we must also know their degree of ionization and this we shall calculate below.

Restriction 2. Charged neutrality must always hold, that is, the entire extrinsic semiconductor must be electrically neutral. We can express this quantitatively referring to Figure 7.6 as,

Total concentration of negative charges = Concentration of conduction electrons in the conduction band + Concentration of electrons attached to acceptor atoms

$$= n_n + n_a^-$$

Similarly,

Total concentration of positive charges = Concentration of holes in valence band + Concentration of positively charged donor ions in the donor levels

$$= n_p + n_d^+$$

Hence

$$n_n + n_a^- = n_p + n_d^+ \qquad (7.36)$$

It is important that the reader realizes that neither the ionized donor atoms (the donor atoms that have donated an electron to the conduction band) nor the ionized acceptor atoms (which really should be called acceptor negative ions) that have accepted an electron from the valence band, thereby leaving a hole there, contribute to the electrical and thermal conduction processes; i.e. they are immobile. Put succinctly, 'conduction only takes place in the conduction and valence bands and not in the impurity levels'.

Actually, equation (7.36) is not quite correct. Because of coulomb repulsion it is unlikely that an acceptor atom will gain two electrons from the valence band even although the Pauli principle allows there to be two provided they are of opposite spins. In practise each acceptor atom usually accepts only one electron. This has the effect of changing the distribution function for electrons in the acceptor levels and that for holes in the valence band. We shall ignore the effect of electron correlation in this book because our analyses are approximate.

In equation (7.36) we have introduced symbols n_d^+ and n_a^-; however, for an extrinsic semiconductor of known doping level we do not know these ionized impurity concentrations. What we do know, however, is the concentration of donor atoms n_d or the concentration of acceptor atoms n_a or both. Our task then is to relate n_d^+ to n_d and n_a^- to n_a and thereby determine the degree of ionization of the impurity atoms. Introduce the symbols

$$n_d^0 = \text{concentration of unionized donor atoms}$$

$$n_a^0 = \text{concentration of unionized acceptor atoms}$$

Then for the case of the donor atoms

$$n_d = n_d^0 + n_d^+ \tag{7.37}$$

We recall that the F.D. distribution function $f_n(\varepsilon)$ gives us the fractional occupancy by an electron of any particular single particle translational energy state. Consider any donor state in the energy gap. The probability that this donor state is occupied by a neutral donor atom of energy ε_d, or that the donor atom has *not* ionized and therefore has *not* given its electron to the conduction band is $f(\varepsilon_d)$, i.e.

$$\frac{n_d^0}{n_d} = f(\varepsilon_d) \tag{7.38}$$

$$= \frac{1}{\exp\left(\dfrac{\varepsilon_d - \varepsilon_F}{kT}\right) + 1}$$

i.e.

$$n_d^0 = \frac{n_d}{\exp\left(\dfrac{\varepsilon_d - \varepsilon_F}{kT}\right) + 1} \tag{7.39}$$

The Quasifree Thermally Perfect Electron Gas

From equation (7.37) we obtain

$$n_d^+ = n_d - n_d^0$$

$$= n_d - \left\{ \frac{n_d}{\exp\left(\frac{\varepsilon_d - \varepsilon_F}{kT}\right) + 1} \right\}$$

$$= \frac{n_d}{1 + \exp\left(-\frac{\varepsilon_d - \varepsilon_F}{kT}\right)} \quad (7.40)$$

whence
Degree of ionization of the impurity donor atoms

$$= \frac{n_d^+}{n_d}$$

$$= \frac{1}{1 + \exp\left(-\frac{\varepsilon_d - \varepsilon_F}{kT}\right)} \quad (7.41)$$

Likewise, we can readily prove that
Degree of ionization of the impurity acceptor atoms

$$= \frac{n_a^-}{n_a}$$

$$= \frac{1}{\exp\left(-\frac{\varepsilon_F - \varepsilon_a}{kT}\right) + 1} \quad (7.42)$$

Unfortunately, these results still contain the Fermi energy which depends upon the level of doping; we can, however, eliminate it thus by considering the donor states. We have the identity

$$-(\varepsilon_d - \varepsilon_F) = \varepsilon_c - \varepsilon_d + \varepsilon_F - \varepsilon_c$$

$$\therefore \exp\left(-\frac{\varepsilon_d - \varepsilon_F}{kT}\right) = \exp\left(\frac{\varepsilon_c - \varepsilon_d}{kT}\right) \cdot \exp\left(-\frac{\varepsilon_c - \varepsilon_F}{kT}\right)$$

But the electron concentration in the conduction band due solely to electrons donated by the donor atoms is, in the saturation region (defined below),

$$n_d^+ = n_c \exp\left(-\frac{\varepsilon_c - \varepsilon_F}{kT}\right) \tag{7.43}$$

[We shall prove this result below—see equations (7.61) and (7.63).]

$$\therefore \qquad \exp\left(-\frac{\varepsilon_d - \varepsilon_F}{kT}\right) = \left[\exp\left(\frac{\varepsilon_c - \varepsilon_d}{kT}\right)\right] \cdot \frac{n_d^+}{n_c}$$

Hence, equation (7.41) becomes

$$\frac{n_d^+}{n_d} = \frac{1}{1 + \left(\dfrac{n_d^+}{n_c}\right) \exp\left(\dfrac{\varepsilon_c - \varepsilon_d}{kT}\right)}$$

This is a quadratic equation in n_d^+ which as the solution

$$n_d^+ = \frac{-n_c + \left[n_c^2 + 4 n_d n_c \exp\left(\dfrac{\varepsilon_c - \varepsilon_d}{kT}\right)\right]^{\frac{1}{2}}}{2 \exp\left(\dfrac{\varepsilon_c - \varepsilon_d}{kT}\right)} \tag{7.44}$$

whence

degree of ionization of impurity donor atoms

$$= \frac{-\left(\dfrac{n_c}{n_d}\right) + \left[\left(\dfrac{n_c}{n_d}\right)^2 + 4\left(\dfrac{n_c}{n_d}\right) \exp\left(\dfrac{\varepsilon_c - \varepsilon_d}{kT}\right)\right]^{\frac{1}{2}}}{2 \exp\left(\dfrac{\varepsilon_c - \varepsilon_d}{kT}\right)} \tag{7.45}$$

This rather awkward expression for the degree of ionization of the donor-impurity atoms contains the donor-atom concentration n_d (the doping level), the temperature and the ionization energy of the donor atoms ($\varepsilon_c - \varepsilon_d$). It is possible to use this equation to draw a family of curves of the degree of ionization versus temperature for known impurity atom concentrations and ionization energy. Such a family of curves is completely analogous to Figure 3.2 for the degree of ionization of a laboratory plasma. We shall not be given this family of curves since a simple numerical calculation using equation (7.45) tells us the really important conclusion that can be drawn from the equation.

Let us evaluate, therefore, the degree of ionization of n-type germanium and silicon at room temperature. Table 7.1 shows that typical ionization energies of donor impurity atoms in germanium are 1.60×10^{-21} J, (0.01 eV) and 8.0×10^{-21} J, (0.05 eV) in silicon. In addition at 300°K the value of kT is 4.14×10^{-21} J (0.026 eV) [see equation (6.18)]. Hence for germanium $\left(\dfrac{\varepsilon_c - \varepsilon_d}{kT}\right)$ has the value of about 0.4 whilst for silicon the value is 2.0.

The Quasifree Thermally Perfect Electron Gas

Now $\exp(0\cdot 4) = 1\cdot 5$ whilst $\exp(2\cdot 0) = 7\cdot 4$. Therefore, let us take an average value of $\exp\left(\dfrac{\varepsilon_c - \varepsilon_d}{kT}\right)$ as $4\cdot 0$. Hence

$$\text{Degree of ionization} \simeq \dfrac{-\dfrac{n_c}{n_d} + \left[\left(\dfrac{n_c}{n_d}\right)^2 + 16\left(\dfrac{n_c}{n_d}\right)\right]^{\frac{1}{2}}}{8}$$

We have, in Section 7.1, discussed the levels of donor impurities that are found in extrinsic semiconductors; these are $2\cdot 50 \times 10^{21} < n_d < 2\cdot 50 \times 10^{25}$ m^{-3}. But $n_c = 2\cdot 5 \times 10^{25}$ per cubic metre from equation (7.16). Hence $1 < \dfrac{n_c}{n_d} < 10^4$. Let us expand the square root by the binomial theorem and obtain

$$\begin{aligned}
\text{Degree of ionization} &= \dfrac{-\left(\dfrac{n_c}{n_d}\right) + \left(\dfrac{n_c}{n_d}\right)\left[1 + \dfrac{16}{2}\left(\dfrac{n_d}{n_c}\right) - \dfrac{16^2}{8}\left(\dfrac{n_d}{n_c}\right)^2 \cdots\right]}{8} \\
&= \dfrac{1}{8}\left[8 - 32\left(\dfrac{n_d}{n_c}\right)\right] \\
&= 1 - 4\left(\dfrac{n_d}{n_c}\right)
\end{aligned}$$

We see that the degree of ionization will exceed 90% at room temperature in both germanium and silicon provided that $\dfrac{n_c}{n_d} > 40$. What this means is that under most conditions of doping the donor atoms in germanium and silicon will be fully ionized at room temperature, i.e.

$$n_d \sim n_d^+ \;\; ; \;\; n_d^0 \sim 0$$

This is a quite generally valid conclusion and perhaps not surprising. After all, the technological importance of extrinsic semiconductors is that they should behave extrinsically at and around room temperature and such extrinsic behaviour will not occur if the impurities are not significantly ionized. It is therefore essential that there be significant ionization of the impurities at and around room temperature, and this we have now shown to be the case.

We are still not yet in a position to calculate the carrier concentrations in the conduction and valence bands of an extrinsic semiconductor. Such a calculation is more complicated than in the intrinsic case; this follows because equations (7.23a) show that in the intrinsic case we do not need to know the position of the Fermi level. However, for the extrinsic case, equations (7.32) and (7.33) show that before the carrier concentrations can be determined we must calculate the Fermi level which depends upon both the level and ionization degree of the dopants. Let us therefore determine the Fermi level in an extrinsic semiconductor.

7.6 The Fermi level in an extrinsic semiconductor

We have already noted in the previous section that the Fermi energy depends upon the amount of doping. We can use equation (7.24) to determine the position of the Fermi level at the absolute zero for any extrinsic semiconductor. At the absolute zero there are no conduction electrons in the conduction band.

$$\varepsilon_{empty} = \varepsilon_c$$

In addition, all donor states are filled with unionized donor atoms so that

$$\varepsilon_{filled} = \varepsilon_d$$

Let us suppose that there are no acceptor atoms present at all. Then

$$\varepsilon_F = \frac{\varepsilon_c + \varepsilon_d}{2} \qquad (7.46)$$

Alternatively, if there are nothing but acceptor atoms present, i.e. no donor impurity atoms at all, then

$$\varepsilon_F = \frac{\varepsilon_a + \varepsilon_V}{2} \qquad (7.47)$$

If, however, as is usually the case, both donor and acceptor atoms are present, but in different amounts, then impurity compensation occurs, there being a partial cancellation or neutralization of the effects of the two types of impurity atoms. Consider an n-type semiconductor and suppose we add acceptor atoms to it. If $n_d > n_a$ even after compensating (the neutralization of some of the donor atoms by the acceptor atoms) so that the semiconductor is still n-type then there will be some empty and some filled donor states so that $\varepsilon_{empty} = \varepsilon_{filled} = \varepsilon_d$

whence

$$\varepsilon_F = \varepsilon_d$$

Similarly for a p-type semiconductor where $n_a > n_d$ then

$$\varepsilon_F = \varepsilon_a$$

However, we know from our discussion of the intrinsic semiconductor that as the temperature rises above absolute zero the Fermi level moves. In the present case this also happens and we can readily locate its position from equations (7.32) and (7.33), viz.

$$\frac{n_n}{n_p} = \frac{n_c}{n_v} \exp\left[\frac{2\varepsilon_F - (\varepsilon_c + \varepsilon_v)}{kT}\right]$$

whence,

$$\varepsilon_F = \frac{\varepsilon_c + \varepsilon_v}{2} + \frac{kT}{2} \ln\left(\frac{n_v}{n_c}\right) + \frac{kT}{2} \ln\left(\frac{n_n}{n_p}\right) \qquad (7.48)$$

But the first two terms on the right-hand side of equation (7.48) are the same as those for the position of the Fermi level of an intrinsic semiconductor of the same material [see equation (7.27)]. Hence

$$\varepsilon_F^{\text{ext}} = \varepsilon_F^{\text{int}} + \frac{kT}{2} \ln\left(\frac{n_n}{n_p}\right) \qquad (7.49)$$

However, we also know that both n_n and n_p must contain the level of doping; how then do we proceed? The answer to this question can be quite involved if we consider the general case of an extrinsic semiconductor with arbitary but different amounts of donor and acceptor impurity atoms. This follows because if we combine equations (7.36) (the charge-neutrality equation) with equations (7.32), (7.33), (7.40) and (7.42) we obtain

$$n_c \exp\left(-\frac{\varepsilon_c - \varepsilon_F}{kT}\right) + \frac{n_a}{1 + \exp\left(-\frac{\varepsilon_F - \varepsilon_a}{kT}\right)}$$

$$= n_v \exp\left(-\frac{\varepsilon_F - \varepsilon_v}{kT}\right) + \frac{n_d}{1 + \exp\left(-\frac{\varepsilon_d - \varepsilon_F}{kT}\right)} \qquad (7.50)$$

This turns out to be a quartic equation in $\exp\frac{\varepsilon_F}{kT}$ which cannot usefully be solved in general terms. However, let us make a very reasonable approximation. Let us assume that there are effectively no acceptor atoms so that we are dealing with an n-type semiconductor. This can be achieved by either omitting them during fabrication by ultrarefining techniques, or by compensation. Then equation (7.36) becomes with $n_a^- = 0$

$$n_n = n_p + n_d^+ \qquad (7.51)$$

However, from equation (7.35)

$$n_p = \frac{n_i^2}{n_n}$$

$$n_n = \frac{n_i^2}{n_n} + n_d^+$$

i.e.

$$n_n^2 - (n_d^+)n_n - n_i^2 = 0$$

This quadratic equation in n_n is readily solved to give

$$n_n = \frac{n_d^+}{2} + \sqrt{n_i^2 + \left(\frac{n_d^+}{2}\right)^2} \qquad (7.52)$$

But equation (7.49), when combined with equation (7.35), gives

$$\varepsilon_F^{\text{ext}} = \varepsilon_F^{\text{int}} + kT \ln\left(\frac{n_n}{n_i}\right) \qquad (7.53)$$

whence

$$\varepsilon_F^{\text{ext}} = \varepsilon_F^{\text{int}} + kT \ln\left\{\left(\frac{n_d^+}{2n_i}\right) \cdot \left[1 + \sqrt{1 + \left(\frac{2n_i}{n_d^+}\right)^2}\right]\right\} \qquad (7.54)$$

which gives the position of the Fermi level for an arbitary doping level. If we analyse this equation we find that as the temperature increases the Fermi level moves downwards from the position given by equation (7.46) *towards the middle of the energy gap* but of course stays above it. Figure 7.7c gives a schematic picture of the position of the Fermi level in an extrinsic semiconductor from the absolute zero as the temperature rises. For 'high' temperatures (and we shall observe below the criterion for just how high the temperature must be), the material becomes intrinsic and the effect of doping on determining the Fermi level becomes negligible. In this case the Fermi level settles at the middle of the energy gap, as we have already discussed in Section 7.4 [see equations (7.27) and (7.28)].

We are now in a position to determine the electron concentration in the conduction band n_n because this is given by equation (7.32) in terms of the Fermi energy whilst equation (7.54) gives the Fermi energy in terms of the level of doping. We have from equation (7.32),

$$n_n = n_c \exp\left(-\frac{\varepsilon_c - \varepsilon_F}{T}\right)$$

$$= n_c \exp\left(-\frac{\varepsilon_c}{kT}\right) \cdot \exp\left(\frac{\varepsilon_F}{kT}\right)$$

The Quasifree Thermally Perfect Electron Gas

But from equation (7.54), on rearranging, the Fermi level is given by

$$\exp\frac{\varepsilon_F}{kT} = \left(\frac{n_d^+}{2n_i}\right)\left[1 + \sqrt{1 + \left(\frac{2n_i}{n_d^+}\right)^2}\right] \cdot \exp\left(\frac{\varepsilon_c + \varepsilon_v}{kT}\right)$$

whence

$$n_n = \left(\frac{n_c n_d^+}{2n_i}\right)\left[1 + \sqrt{1 + \left(\frac{2n_i}{n_d^+}\right)^2}\right] \exp\left(-\frac{\varepsilon_G}{2kT}\right) \quad (7.55)$$

We must now express n_d^+ in terms of n_d using equation (7.44). Clearly this is going to give messy algebra. Rather than obscure the analysis by such algebra, let us make a further approximation to the analysis; we shall suppose that the donor atom impurity doping is sufficiently great to put a number of extrinsic electrons in the conduction band—so many, in fact, that *we can ignore intrinsic effects altogether*. This means that effectively all the electrons in the conduction band are extrinsic conduction electrons, i.e. n_n is 'large' and much greater than n_i. But from equation (7.35), $n_p = \frac{n_i^2}{n_n}$ so that n_p is 'small'. We shall put $n_p = 0$. In this approximation, equation (7.52) becomes

$$n_n \sim \left(\frac{n_d^+}{2}\right) + \sqrt{\left(\frac{n_d^+}{2}\right)^2}$$

since

$$n_i \ll n_d^+$$

i.e.

$$n_n \sim n_d^+ \quad (7.56)$$

(This equation could have been deduced directly from equation (7.36) by putting

$$n_a^- \equiv 0 \quad ; \quad n_p \sim 0)$$

Therefore

$$n_c \exp\left(-\frac{\varepsilon_c - \varepsilon_F}{kT}\right) \sim \frac{n_d}{1 + \exp\left(-\frac{\varepsilon_d - \varepsilon_F}{kT}\right)} \quad (7.57)$$

using equation (7.40).

If we rearrange equation (7.57) we find that it gives the following quadratic equation

$$\left[n_c \exp\left(-\frac{\varepsilon_c + \varepsilon_d}{kT}\right)\right]\left(\exp\frac{\varepsilon_F}{kT}\right)^2 + \left[n_c \exp\left(-\frac{\varepsilon_c}{kT}\right)\right]\left(\exp\frac{\varepsilon_F}{kT}\right) - n_d = 0$$

(7.58)

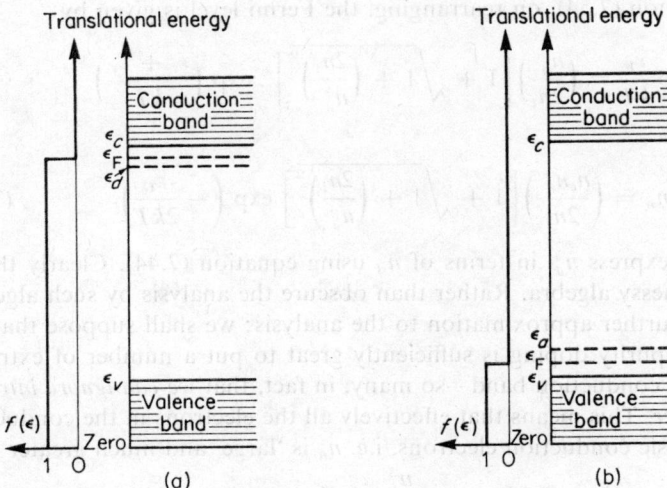

Figure 7.7. The extrinsic semiconductor

Figure 7.7(a). An n-type extrinsic semiconductor at the absolute zero. The F.D. probability distribution function $f(\varepsilon)$ is shown which has a value of unity up to the Fermi level and zero thereafter in the conduction band which is empty. No electrons are shown occupying the valence band which is in fact full and there are no holes in this band. No donor atoms are ionized.

Figure 7.7(b). A p-type extrinsic semiconductor at the absolute zero. The F.D. probability distribution function $f(\varepsilon)$ is shown which has a value of unity up to the Fermi level and zero thereafter in the conduction band which is empty. No electrons are shown occupying the valence band which is in fact full and there are no holes in this band. No acceptor atoms are 'ionized', i.e. none have gained an additional electron from the valence band.

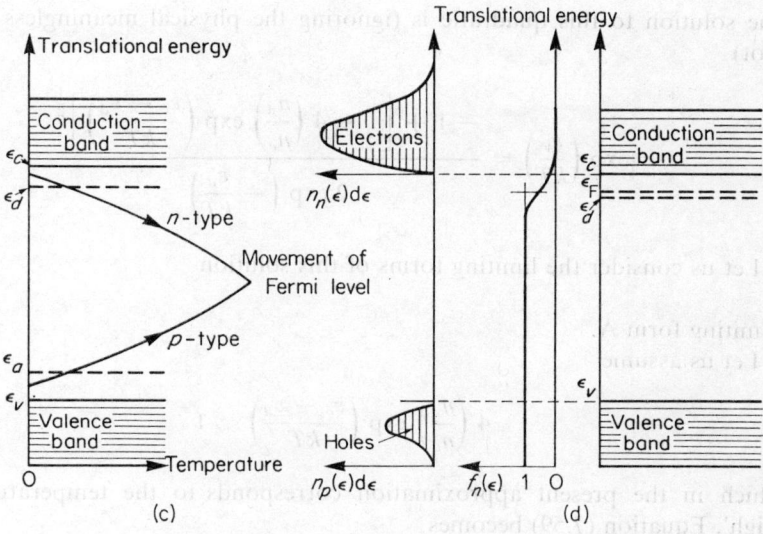

Figure 7.7(c). A schematic representation of the movement of the Fermi level with increase of temperature for an n-type and a p-type extrinsic semiconductor. No electrons, holes or ionized impurities are shown although these are present in their respective energy levels.

Figure 7.7(d) An n-type extrinsic semiconductor at some finite non-zero temperature. The left-hand graph shows the carrier concentrations of electrons in the conduction band and holes in the valence band as a function of the carrier energy. Note the excess in a total concentration of electrons over holes. The right-hand graph shows the F.D. electron probability distribution function $f(\varepsilon)$ which now has a non-zero value in the conduction band because this band is partially occupied by electrons thermally excited from the donor atom levels. No electrons, holes or ionized impurities are shown although these are present in their respective energy levels.

The solution to this quadratic is (ignoring the physical meaningless negative root)

$$\exp\left(\frac{\varepsilon_F}{kT}\right) = \frac{-1 + \left[1 + 4\left(\frac{n_d}{n_c}\right)\exp\left(\frac{\varepsilon_c - \varepsilon_d}{kT}\right)\right]^{\frac{1}{2}}}{2\exp\left(-\frac{\varepsilon_d}{kT}\right)} \tag{7.59}$$

Let us consider the limiting forms of this solution

Limiting form A.
Let us assume

$$4\left(\frac{n_d}{n_c}\right)\exp\left(\frac{\varepsilon_c - \varepsilon_d}{kT}\right) \ll 1 \tag{7.60}$$

which in the present approximation corresponds to the temperature being 'high'. Equation (7.59) becomes

$$\exp\left(\frac{\varepsilon_F}{kT}\right) \sim \frac{-1 + \left[1 + 2\left(\frac{n_d}{n_c}\right)\exp\left(\frac{\varepsilon_c - \varepsilon_d}{kT}\right)\right]}{2\exp\left(-\frac{\varepsilon_d}{kT}\right)}$$

i.e.

$$\exp\left(\frac{\varepsilon_F}{kT}\right) \sim \left(\frac{n_d}{n_c}\right) \cdot \exp\frac{\varepsilon_c}{kT}$$

i.e.

$$n_d \sim n_c \exp\left(-\frac{\varepsilon_c - \varepsilon_F}{kT}\right) \tag{7.61}$$

The right-hand side of this equation is identical with the right-hand side of equation (7.32) so that their left-hand sides must also be identical, i.e.

$$n_n = n_d = \text{constant} \tag{7.62}$$

The only conclusion that we can draw from equations (7.56) and (7.62) is that all the donor levels must be ionized, i.e.

$$n_n = n_d = n_d^+ \tag{7.63}$$

Hence the electron concentration in the conduction band is a *constant* and in fact a maximum because there are no more donor atoms left to ionize to give any more electrons to the conduction band—with the proviso that this result

The Quasifree Thermally Perfect Electron Gas

only holds for extrinsic semiconductors such that all intrinsic effects are negligible. This region where intrinsic effects are negligible and the electron concentration in the conduction band is a constant, maximum value, is known as the saturation region.

Limiting Form B.

Let us assume that

$$4 \left(\frac{n_d}{n_c}\right) \exp\left(\frac{\varepsilon_c - \varepsilon_d}{kT}\right) \gg 1 \tag{7.64}$$

which in the present approximation corresponds to the temperature being low. Then equation (7.59) becomes

$$\exp\left(\frac{\varepsilon_F}{kT}\right) = \left(\sqrt{\frac{n_d}{n_c}}\right) \cdot \exp\left(\frac{\varepsilon_c + \varepsilon_d}{kT}\right)$$

But from equation (7.32)

$$n_n = n_c \exp\left(-\frac{\varepsilon_c - \varepsilon_F}{kT}\right)$$

$$= n_c \exp\left(-\frac{\varepsilon_c}{kT}\right) \cdot \left(\sqrt{\frac{n_d}{n_c}}\right) \cdot \exp\frac{\varepsilon_c + \varepsilon_d}{2kT}$$

$$= (\sqrt{n_d n_c}) \cdot \exp\left(-\frac{\varepsilon_c - \varepsilon_d}{2kT}\right) \tag{7.65}$$

Hence the electron concentration in the conduction band is directly proportional to the square root of the donor impurity-atom concentration and exponentially proportional to the temperature of the extrinsic semiconductor.

Figure 7.8 is a schematic representation of the electron concentration in the conduction band as a function of temperature for an extrinsic semiconductor containing donor impurity atoms only. As the temperature is raised from the absolute zero, conduction electrons are thermally excited from the donor levels into the conduction band and as the temperature increases so more and more of these donor atom electrons are thermally excited into the conduction band. Consequently the electron concentration in the conduction band rapidly increases (exponentially with temperature). However, beyond a certain point no additional electrons will be available from the donor atoms because each donor atom is now ionized and therefore all donor atoms have contributed their electrons to the conduction band. A further increase of temperature will result in intrinsic effects becoming important as electrons start to enter the conduction band by direct thermal excitation across the energy gap from the valence band in ever-increasing numbers. Since there are a vast number of electrons in the valence band to act as an electron 'source' eventually the electron concentration in the conduction band becomes swamped by those electrons that have come directly from the valence band rather than from the donor impurity levels. This means that extrinsic effects become less and less noticeable whilst intrinsic effects become more and more dominant until eventually the vast bulk of the electrons in the conduction band have come direct from the valence band; thus the semiconductor is now intrinsic in behaviour. When

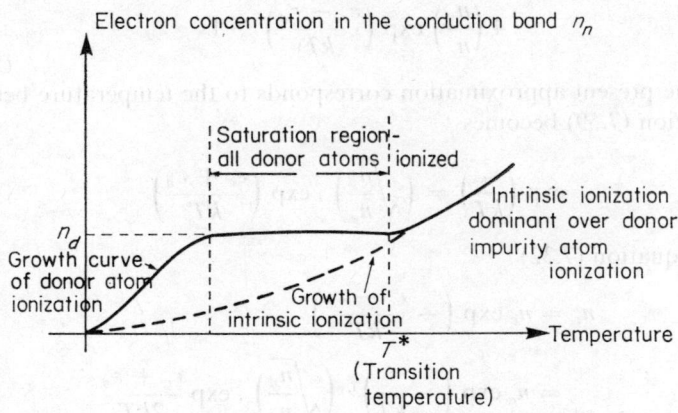

Figure 7.8 Schematic representation of the electron concentration in the conduction band as a function of temperature for an *n*-type extrinsic semiconductor. The three regions are clearly shown where the donor impurity atom ionization grows from zero to a maximum corresponding to total ionization followed by a region of constant total donor impurity atom ionization, i.e. the saturation region followed finally by the post transition temperature region where intrinsic effects dominate the carrier concentration and where the effect of donor impurity atom ionization becomes negligible.

this happens the electron concentration in the conduction band is given by the intrinsic electron concentration formula, equation (7.23a) viz.

$$n_n = n_i = n_c \exp\left(-\frac{\varepsilon_G}{2kT}\right)$$

In equation (7.45) we found that, at room temperature, the degree of ionization of the donor impurity atoms in both germanium and silicon is over 90% so that we have some idea of the temperature at which the transition from the low-temperature range of equation (7.65) to the saturation range given by equation (7.62) takes place. Let us see if we can estimate the temperature T^* at which the transition between equation (7.63) and (7.23a) occurs, that is, the temperature at which the increasing intrinsic electron concentration in the conduction band becomes comparable with the constant, maximum extrinsic

The Quasifree Thermally Perfect Electron Gas

electron concentration in the conduction band in the saturation range. Clearly, this will be given by a simultaneous solution of the two equations

$$n_n = n_d$$
$$n_n = n_i$$

i.e.

$$n_d = n_c \exp\left(-\frac{\varepsilon_G}{2kT^*}\right)$$

i.e.

$$T^* = \frac{\varepsilon_G}{2k \ln\left(\frac{n_c}{n_d}\right)} \tag{7.66}$$

Therefore the transition temperature T^* depends upon the width of the energy gap and the smaller this gap the lower the transition temperature. Let us divide equation (7.66) by room temperature 300°K and get

$$\frac{T^*}{T_{300}} = \frac{\varepsilon_G}{2kT_{300} \ln\left(\frac{n_c}{n_d}\right)}$$

But from equation (6.18)

$$kT_{300} = 4 \cdot 14 \times 10^{-21} \text{ J } (0 \cdot 026 \text{ eV})$$

whence

$$\frac{T^*}{T_{300}} = \frac{\varepsilon_G}{8 \cdot 28 \times 10^{-21} \ln\left(\frac{n_c}{n_d}\right)} \tag{7.67}$$

Clearly this equation gives the transition temperature for a semiconductor of known energy gap and known impurity atom concentration. Let us evaluate it for silicon and germanium, using the values of the energy gap given in Table 7.2. For germanium

$$\frac{T^*}{T_{300}} = \frac{1 \cdot 04 \times 10^{-19}}{8 \cdot 28 \times 10^{-21} \ln\left(\frac{n_c}{n_d}\right)} = \frac{12 \cdot 5}{\ln\left(\frac{n_c}{n_d}\right)}$$

For silicon

$$\frac{T^*}{T_{300}} = \frac{1 \cdot 84 \times 10^{-19}}{8 \cdot 28 \times 10^{-21} \ln\left(\frac{n_c}{n_d}\right)} = \frac{22 \cdot 2}{\ln\left(\frac{n_c}{n_d}\right)}$$

Consider lightly doped germanium and silicon such that $n_d \sim 2\cdot5 \times 10^{21}$ m^{-3}. The value of n_c is given by equation (7.16) so that

$$\ln\left(\frac{n_c}{n_d}\right) = \ln\left(\frac{2\cdot5 \times 10^{25}}{2\cdot5 \times 10^{21}}\right) = \ln 10^4 = 9\cdot2$$

Hence

$$T^*_{\text{Ge}} = 408°\text{K} = 135°\text{C}$$

and

$$T^*_{\text{Si}} = 724°\text{K} = 451°\text{C}$$

In some respects this calculation is misleading. The reader will well know that the really important feature of of semiconductor materials is the p–n junction (which we shall discuss later) rather than the bulk material itself. The actual working volume of a transistor, for example, seldom exceeds 10^{-9} cubic metres even for high-power applications, so that finite thermal energy production (joule dissipation) is occuring in a very small volume and this results in there being a power dissipation problem arising in almost all transistor applications. Now the electrical characteristics of the junction depend upon the operating temperature and this in turn depends upon the power dissipated in the device. If this power is large then the temperature of the device will rise and, as the above calculation has shown, care must be taken to see that the above transition temperatures are not reached during operation. Most commercial transistors are supplied with both upper and lower storage and operating temperatures prescribed in addition to the maximum power that the device can dissipate. The lower storage temperature is quoted so that mechanical fracture damage to the transistor is avoided due to elastic strain occurring on differential thermal expansion of the joints between the metallic solders and the semiconductor materials; the upper storage temperature is quoted so that these solders are never melted or suffer chemical oxidation. In practise the minimum storage temperatures range from about $-50°$C to $-75°$C whilst the upper storage temperatures are about $100°$C for germanium and $300°$C for silicon. The operating temperatures are quoted because if the temperature falls too low the impurity ionization falls off whilst if the temperature is too high the materials become intrinsic and the p–n junction loses its distinctive features. Generally speaking, the operating temperatures lie within the storage temperatures and in practise germanium should not be operated much above $85°$C, say, whilst silicon should not be operated above $200°$C. Provided that these upper operating temperature limits are not exceeded, either semiconductor will be working in the saturation region where the electron concentration in the conduction band is constant, independent of temperature from room temperature upwards,

The Quasifree Thermally Perfect Electron Gas

thereby giving electrically stable devices. We shall return to this question of power dissipation and operating temperatures at the end of Section 7.8 after we have discussed the thermal conductivity of semiconductors.

7.7 Carrier mobility and carrier scattering processes in a semiconductor

Many features of this section are similar to those introduced in Chapter 6, where we discussed metals. Experimentally it is observed that at low electric field strengths, less than say 10^8 volts per metre, semiconductors obey Ohm's law, viz. equation (6.2)

$$j = \sigma E_F$$

For higher field strengths (such as in fact may occur at a p–n junction) departures from Ohm's law can occur. However, in the range of field strengths where Ohm's law is obeyed we can define an electrical conductivity

$$\sigma = \frac{j}{E_F}$$

Analogously to equation (6.7) where we introduced the drift velocity we can write, for any charge carrier α

$$j = n_\alpha (Z_\alpha e) v_\alpha$$

where

$n_\alpha = $ number of charged carriers per unit volume of the semiconductor
$Z_\alpha = -1$ for electrons and $+1$ for holes
$v_\alpha = $ drift velocity of the charge carrier.

We also define
(Drift) mobility of a charge carrier

$$= \mu_\alpha$$
$$= \frac{|v_\alpha|}{E_F} \tag{7.68}$$

We note that the units of mobility are $m^2/\text{V s}$.

Of course electrons and holes drift in different directions under the influence of the field E_F but in both cases their mobilities are *always positive*, i.e.

$$\left. \begin{array}{l} v_n = -\mu_n E_F \\ v_p = \mu_p E_F \end{array} \right\} \tag{7.69}$$

Equations (6.7) and (6.17) when combined give

$$(-e)v_e = \frac{e^2 \cdot E_F \cdot \tau}{m_e^*}$$

i.e.

$$\frac{v_e}{E_F} = \frac{(-e)\tau}{m_e^*}$$

Analogously to this equation we can introduce two relaxation times for electrons and holes such that

$$\left.\begin{array}{l} \mu_n = -\dfrac{v_n}{E_F} = -\dfrac{(-e)\tau_n}{m_n^*} = \dfrac{e\tau_n}{m_n^*} \\[2mm] \mu_p = \dfrac{v_p}{E_F} = \dfrac{e\tau_p}{m_p^*} \end{array}\right\} \qquad (7.70)$$

where m_p^* is the effective mass of a hole.

We can express the electrical conductivity in terms of mobility; let the total electrical current density in the semiconductor be j then

$$\begin{aligned} j &= j_n + j_p \\ &= n_n(-e)v_n + n_p(e)v_p \\ &= n_n e \mu_n E_F + n_p e \mu_p E_F \\ &= (n_n e \mu_n + n_p e \mu_p) E_F \\ &= \sigma E_F \end{aligned}$$

whence

$$\sigma = e(n_n \mu_n + n_p \mu_p) \qquad (7.71)$$

At this point it is useful to note once again the basic difference between the electrical conductivity of a metal and that of a semiconductor. For a metal equations (6.17) and (6.24) show that

$$\sigma = \frac{n_e \cdot e^2 \cdot \tau}{m_e^*} = \frac{n_e \cdot e^2}{m_e^*} \cdot \frac{l_{\text{Fermi}}}{v_{\text{Fermi}}}$$

In metals n_e is the number of valence electrons per unit volume which is a *constant*: also from equation (6.20) v_{Fermi} is a constant so that

$$\sigma \propto l_{\text{Fermi}}$$

We also found in Section 6.4 that l_{Fermi} decreases with increasing temperature and we concluded that the electrical conductivity of a metal *decreased* with *increasing* temperature. However, for both intrinsic and extrinsic semiconductors

The Quasifree Thermally Perfect Electron Gas

the charge-carrier concentrations n_n and n_p *are not constant*; on the contrary, as we have seen, they are very strongly dependent on the temperature and they increase exceedingly rapidly with increasing temperature. Hence the electrical conductivities of intrinsic and extrinsic semiconductors (and insulators as well!) *increase* with increasing temperature.

Another essential difference between a metal and a semiconductor is that the electrical conductivity of a metal increases with its purity or lack of foreign atoms, whereas the electrical conductivity of a semiconductor decreases with its purity—in fact, extrinsic semiconductors are totally dependent on ionized impurities for their important electrical conduction properties at and around room temperature, provided of course they are outside the intrinsic region. Yet another point of difference between a metal and a semiconductor is that when we consider the latter we must be careful how we identify the collisional relaxation times introduced in equation (7.70). In the case of a metal, equation (6.19) showed that

$$\varepsilon_F \gg \bar{\varepsilon}$$

i.e. the Fermi energy is very much greater than the average thermal kinetic energy. The result of this is that the collisional relaxation time is

$$\tau_{\text{Fermi}} = \frac{l_{\text{Fermi}}}{v_{\text{Fermi}}}$$

where v_{Fermi} is a constant. However, in the case of a semiconductor the situation is very different. Firstly, electrons in the conduction band (and holes in the valence band) do *not* have some 'zero-point' kinetic energy, that is, kinetic energy at the absolute zero of temperature, as do essentially all electrons in a metal (see Sections 4.3 and 6.3). Rather, at the absolute zero in the case of a semiconductor electrons have *zero* kinetic energy because there are *none* in the conduction band whilst of course there will be no holes in the valence band either; this point was stressed in Section 7.1. Hence, *all* electrons in the conduction band and holes in the valence band have kinetic energies that fall to zero at the absolute zero, which means that the kinetic energies of electrons and holes at all finite non-zero temperatures are truly *thermal* kinetic energies or energies due to random thermal motion. Furthermore, in the classical approximation that we have used in Section 7.3 we showed that the distribution of electrons in the conduction band and holes in the valence band is essentially an M.B. distribution. Equation (3.3b) showed that the average thermal kinetic energy of a particle in an M.B. gas is

$$\bar{\varepsilon} = \tfrac{3}{2}kT$$

But, following equation (6.6), the average thermal kinetic energy is also

$\frac{1}{2}m^*\overline{v^2}$. Hence, on equating these two expressions for the average thermal kinetic energy we get

$$\overline{v_{th}} = \sqrt{\overline{v^2}} = \sqrt{\frac{3kT}{m^*}}$$

where $\overline{v_{th}}$ is the root mean square thermal speed of a particle. Now in general $l = v\tau$ and for a metal, we wrote this as equation (6.24) viz.

$$l_{Fermi} = v_{Fermi} \cdot \tau_{Fermi} \quad \text{with } v_{Fermi} = \text{constant}$$

However, for a semiconductor, since we are considering a true thermal situation we have

$$l = \overline{v_{th}} \cdot \tau \tag{7.72}$$

where $\overline{v_{th}}$ is *not* a constant, rather it is a function of the square root of the temperature. Hence

$$\tau = \frac{l}{\sqrt{\frac{3kT}{m^*}}} \tag{7.73}$$

Using equation (7.70) we get

$$\mu_\alpha = \frac{Z_\alpha e_\alpha \tau_\alpha}{m_\alpha^*}$$

$$= \frac{Z_\alpha e_\alpha}{m_\alpha^*} \cdot \frac{l_\alpha}{\sqrt{\frac{3kT}{m_\alpha^*}}}$$

$$= \frac{Z_\alpha e_\alpha}{\sqrt{3kTm_\alpha^*}} \cdot l_\alpha \tag{7.74}$$

i.e.

$$\mu_\alpha \propto \frac{l_\alpha}{\sqrt{T}} \tag{7.75}$$

Unlike our analysis in Chapter 6 we shall discuss scattering mechanisms in a semiconductor, not in terms of the temperature-dependence of the mean free path, but in terms of the temperature-dependence of the mobility μ_α. The above analysis assumes (as did the analysis in Chapter 6 for a metal) that the drift velocity of the carriers in the externally applied electric field was smaller than the root mean square thermal speed (see discussion at the end of Section 6.3).

i.e.

$$|v_\alpha| < \overline{v_{th}}$$

If the electric field, however, is increased to above 10^8 V/m then ultimately

The Quasifree Thermally Perfect Electron Gas

the drift velocity reached by the carriers becomes comparable with, or even greater than, the thermal speed. If this happens we can readily show that the drift velocity becomes proportional to the square root of the applied field and this means that the carriers 'heat up' or gain a higher energy from the applied field than they lose in scattering collisions. This phenomenon occurs in semiconductors but not in metals. We shall not, however, discuss it further.

The mobility of a carrier is determined by the various scattering processes that they suffer as they move under the influence of the externally applied field. In a semiconductor we consider

1. Carrier–phonon scattering collisions.
2. Carrier–ionized impurity atom scattering collisions.
3. Carrier–neutral atom scattering collisions.

We found in Chapter 6 that the value of l_{Fermi} in a metal depended upon some simple power law of temperature; the same conclusion holds for carrier mobility in a semiconductor. However, we must remind ourselves that whereas in a metal only those special valence electrons at or around the Fermi energy, that is, the conduction electrons suffer scattering (because electrons lying 'deeper' could not find empty translational energy states to move into), in the case of the semiconductor because the carriers are not numerous (in fact we have used M.B. statistics to discuss them and this requires that there be 'few' particles to distribute over 'many' translational energy states—see Sections 2.9 and 2.13); this means that for electrons in the conduction band and holes in the valence band there will likely always be some empty allowed translational energy states close by into which a carrier from a neighbouring energy state can be scattered. What this implies is that in a semiconductor we must consider the effect of scattering collisions with a range of carrier translational energies rather than just one translational energy (the Fermi energy) for the case of the conduction electrons in a metal.

Our discussion of the temperature-dependence of the mobility will be entirely qualitative because the quantal analysis to find the scattering collision cross-section (see Section 6.4) that determines the mobility is beyond the scope of this book.

Let us consider lattice scattering or carrier-phonon scattering. Equation (6.32a) shows that in a metal the mean free path is for $T > \theta_\text{D}$

$$l_{\text{Fermi}} \propto \frac{\theta_\text{D}^2}{T}$$

If it can be assumed that this result is also true for the carriers in a semiconductor then we can write from equation (7.75)

$$\mu_\alpha \propto \frac{\theta_\text{D}^2}{T^{\frac{3}{2}}} \qquad (7.76)$$

In chemically and physically pure semiconductors (which therefore must be intrinsic) lattice scattering is the dominant 'resistance to motion' for $T > \theta_D$ and in this case

$$\mu_\alpha \propto T^{-\frac{3}{2}}$$

Clearly this behaviour cannot apply at very low temperatures $T \ll \theta_D$ because the phonon density will have become too low. What then scatters the carriers in an intrinsic semiconductor at very low temperatures? If exceedingly small amounts of (residual) impurities are present (and our discussion in Section 7.1 showed that the minimum impurity was still about one part in 10^{10}) then carrier-residual impurity atom scattering will be the dominant process just as it was in the case of electrical conductivity of a metal at very low temperatures where we introduced the idea of the residual resistivity in equation (6.36). In addition, a calculation based on equation (7.45) shows that at these very low temperatures the residual impurities will not be ionized $kT \ll (\varepsilon_c - \varepsilon_d)$ and this of course explains why the semiconductor is intrinsic. Hence we are dealing with carrier-*neutral* residual atom impurity scattering. Analysis shows that in this case the mobility is

$$\mu_\alpha \propto (n_d)^{-1} \qquad (7.77)$$

i.e. it is independent of temperature.

In the intermediate temperature range between $T > \theta_D$ and $T \ll \theta_D$ these residual impurities will begin to ionize. If the semiconductor is very pure and therefore intrinsic then carrier-ionized impurity scattering will not be important until the phonon density has fallen to a low value. Analysis shows that for carrier-ionized impurity scattering

$$\mu_\alpha \propto T^{\frac{3}{2}} \qquad (7.78)$$

which gives the rather surprising result that the mobility increases with increasing temperature. The physical explanation of this is related to the fact that carrier-ionized impurity atom scattering is a type of Debye–Hückel screened-coulomb encounter between charged particles and we get small-angle scattering collisions (a similar situation was referred to in Section 6.9) which do not really significantly impede the motion of the carriers. (Small-angle scattering between charged particles is also the essence of the problem of determining the electrical conductivity of a highly ionized laboratory plasma which, however, we shall not discuss in this book.) Figure 7.9 shows a schematic picture of the variation of the mobility of an intrinsic semiconductor with temperature. In practise there is a considerable overlap of the scattering regions.

If, however, the semiconductor is extrinsic, then scattering of the carriers by impurities will obviously become increasingly important as the doping increases. Figure 7.9 also shows the variation of the mobility of an extrinsic semiconductor with temperature. At any particular temperature the carrier mobility will depend upon the concentration of impurities and their degree of ionization and we cannot give simple equations like equations (7.76), (7.77) and (7.78) which apply for intrinsic semiconductors. The *intrinsic* maximum in the mobility curve occurs for both germanium and silicon at about 80°K but moves towards higher temperatures as the impurity level increases; for heavily doped silicon and germanium the maximum, which is not very pronounced, occurs at about room temperature. This represents another significant difference between a metal and an extrinsic semiconductor. We have seen in

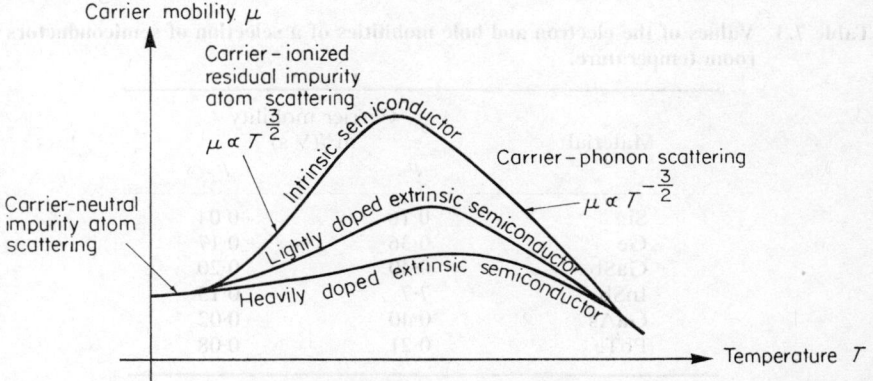

Figure 7.9. Schematic representation of the variation of the carrier mobility for both intrinsic and extrinsic semiconductors as a function of temperature. The three scattering regions are shown and it should be noted that the maximum in the mobility curves decreases in magnitude and moves to higher temperatures as the doping level increases.

Chapter 6 that in a metal, scattering by impurities is only important at low temperatures (the residual electrical resistivity of a metal was only encountered at temperatures less than about $0.03\,\theta_D$—see Section 6.4). However, in the case of a heavily doped semiconductor, ionized impurity atom scattering can prevail right up to room temperature.

Table 7.3 gives values of carrier mobility at 300°K.

Although we did not apply the concept of mobility in Chapter 6 in our discussion of metals, it is of some interest to calculate the mobility of a typical metal. From equation (6.17) we have

$$\sigma = \frac{n_e \cdot e^2 \cdot \tau}{m_e^*}$$

and from equation (7.70), adapting the equation to apply to a metal we have

$$\mu_e = \frac{(-e) \cdot \tau}{m_e^*}$$

$$\sigma = n_e \cdot (-e) \cdot \mu_e$$

i.e.

$$\mu_e = \frac{1}{\rho_e \cdot n_e \cdot e}$$

dropping the minus sign. But from Chapter 6 we found, for the case of silver, that the valence-electron concentration was $5 \cdot 9 \times 10^{28}$ electrons per cubic metre.

Table 7.3 Values of the electron and hole mobilities of a selection of semiconductors at room temperature.

Material	Carrier mobility m²/(V s)	
	μ_n	μ_p
Si	0·16	0·04
Ge	0·36	0·17
GaSb	0·40	0·20
InSb	7·7	0·13
GaAs	0·40	0·02
PbTe	0·21	0·08

Also at room temperature Table 6.1 shows that the electrical resistivity is $1 \cdot 6 \times 10^{-8}$ Ωm. Hence

$$\mu_e = \frac{1}{1 \cdot 6 \times 10^{-8} \times 5 \cdot 9 \times 10^{28} \times 1 \cdot 6 \times 10^{-19}}$$
$$= 6 \cdot 6 \times 10^{-3} \, \text{m}^2/\text{V s}$$

If we compare this value with the values given in Table 7.3 we see that the ratio of the mobility of a semiconductor to that of a metal is about one hundred to one. The larger value of the mobility of a semiconductor compared with that of a metal is primarily due to the larger value of the collisional relaxation time τ of a carrier in a semiconductor. This is a consequence of the fact that the carrier concentration in a semiconductor is about 10^{20} electrons per cubic metre [see equation (7.30)] which should be compared with the 10^{28} valence electrons per cubic metre in a metal so that on average a carrier will travel further in a semiconductor than it will in a metal. Also the carrier speed in a semiconductor, viz. v_{th} is much less than that in a metal, viz. v_{Fermi}.

We can use the data given to prove that it is highly unlikely that an externally applied electric field, as opposed to an externally applied temperature gradient, can raise electrons from the valence band into the conduction band. Consider an electron at the top of the valence band when an external field of magnitude E_F is applied to the semiconductor

$$\text{Energy gained in a mean free path} = (-e) \cdot E_F \cdot l_n$$

But from equation (7.74)

$$l_n = \frac{\sqrt{3kT \cdot m_n^*} \cdot \mu_n}{(-e)}$$

∴

$$\text{Energy gained in a mean free path} = \sqrt{3kT \cdot m_n^*} \cdot \mu_n \cdot E_F$$

Consider the case of silicon at room temperature subjected to an external electric field of 10^4 V/m.

Energy gained per mean free path

$$= (3 \times 1\cdot38 \times 10^{-23} \times 300 \times 9\cdot11 \times 10^{-31})^{\frac{1}{2}} \times 0\cdot16 \times 10^4$$
$$= 1\cdot70 \times 10^{-22} \text{ J } (0\cdot011 \text{ eV}) \qquad (7.79)$$

The energy gap in silicon is from Table 7.2, $1\cdot84 \times 10^{-19}$ J (1·15 eV). Clearly for the case of *intrinsic* silicon the energy gained per free path in the external electric field is only about one hundredth of the energy required by an electron to cross the energy gap; hence there will be effectively no direct transfer of electrons from the valence band to the conduction band due to the action of the electric field. This conclusion is also the same for *extrinsic* silicon because we have seen from Table 7.1 that the ionization potential of n-type silicon is about $8\cdot0 \times 10^{-21}$ J (0·05 eV) so once again there will be effectively no transfer of electrons from the donor atoms to the conduction band due to the action of the electric field. The only common example where high electric fields can cause changes in constant temperature semiconductors is the Zener effect which however we shall not discuss here.

7.8 The electrical and thermal conductivity of a semiconductor

Having discussed mobility we are now in a position to determine the electrical conductivity of a semiconductor.

Equation (7.71) is

$$\sigma = e(n_n \mu_n + n_p \mu_p)$$

We see therefore that any discussion of the temperature dependence of the electrical conductivity will require a discussion of the temperature dependence of *both* the carrier concentration *and* the carrier mobility [we recall that in a metal the carrier (i.e. valence electron) concentration is *not* a function of temperature].

Consider firstly the intrinsic semiconductor. Here, from equation (7.23a) we have

$$n_n = n_p = n_i = n_c \exp\left(-\frac{\varepsilon_G}{2kT}\right) = n_V \exp\left(-\frac{\varepsilon_G}{2kT}\right)$$

Substituting into equation (7.71) we get

$$\sigma = en_i(\mu_n + \mu_p) \tag{7.80}$$

Now from equations (7.15), (7.17) and (7.18) we have both n_c and n_V proportional to $T^{\frac{3}{2}}$. Hence

$$n_i \propto T^{\frac{3}{2}} \exp\left(-\frac{\varepsilon_G}{2kT}\right)$$

whence

$$\sigma \propto T^{\frac{3}{2}}(\mu_n + \mu_p) \exp\left(-\frac{\varepsilon_G}{2kT}\right) \tag{7.81}$$

Consider the high-temperature limit where $T > \theta_D$. From equation (7.76) we have that both μ_n and μ_p are proportional to $T^{-\frac{3}{2}}$ whence

$$\sigma \propto \exp\left(-\frac{\varepsilon_G}{2kT}\right) \tag{7.82}$$

Thus the electrical conductivity of an intrinsic semiconductor increases exponentially with increasing temperature and decreasing energy gap. Equation (7.82) is also

$$\ln \sigma = -\left(\frac{\varepsilon_G}{2kT}\right)$$

so that a graphical plot of $\ln \sigma$ against T^{-1} should be a straight line of gradient $-\dfrac{\varepsilon_G}{2k}$. This represents the basis of an experimental determination of the energy gap of an intrinsic semiconductor from measurement of the intrinsic conductivity. An interesting and important practical application of equation (7.82) is the thermistor (thermally sensitive resistor). This uses the relatively low resistance but high temperature coefficient of electrical resistance of an intrinsic semiconductor. We have, following equation (6.35),

Temperature coefficient of electrical resistance

$$= \frac{1}{\rho}\frac{d\rho}{dT}$$

$$= -\frac{\varepsilon_G}{2kT^2} \tag{7.83}$$

The Quasifree Thermally Perfect Electron Gas

by inverting and differentiating equation (7.82). We note that the thermistor has a negative temperature coefficient of electrical resistance and for certain semiconductors this can be numerically large, about 0·04/deg.C. If we put a thermistor in a metallic circuit which has a positive temperature coefficient of electrical resistance (as all metals do—see Table 6.1) then with proper adjustment of the thermistor behaviour we can obtain a circuit whose overall behaviour is independent of temperature over a wide range—say 100 deg C. The semiconductor thermistor is used for industrial temperature measurement and control and also, when used as a bolometer, for measuring infrared and microwave power levels. Another application of equation (7.82) involving a reduction of electrical resistivity for an increase of temperature is the varistor. This is a non-ohmic device which utilizes internal joule heating in the semiconductor (usually SiC) to reduce the electrical resistance of the varistor. It is usually connected in parallel across electrical equipment and under normal conditions draws very little current. However, under surge conditions its resistance drops rapidly and it effectively carries the surge current.

Turning to the low-temperature limit $T \ll \theta_D$ for the intrinsic semiconductor we have, from equation (7.77), the fact that mobility is independent of temperature so that equation (7.81) becomes

$$\sigma \propto T^{\frac{3}{2}} \exp\left(-\frac{\varepsilon_G}{2kT}\right) \tag{7.84}$$

Usually the $T^{\frac{3}{2}}$ is swamped by the exponential temperature variation so that in practise we obtain much the same behaviour as equation (7.82).

This conclusion likewise holds true for the intermediate temperature range of an intrinsic semiconductor because the mobility is likely dominated by electron-ionized residual impurity atom scattering which is given by equation (7.78). We therefore conclude that in most cases the electrical conductivity of an intrinsic semiconductor is given by equation (7.82) over the complete temperature range.

Let us now consider the extrinsic semiconductor. We shall discuss the n-type since we have developed the appropriate equations for such an extrinsic semiconductor. Our conclusions, however, hold for p-type materials as well.

We have shown that for a small level of doping both the high and low-temperature carrier concentrations are given by equations (7.62) and (7.65) respectively, i.e. in the high-temperature case $n_n = n_d$ = constant with all donor impurities ionized, whilst in the low-temperature case we have

$$n_n = (\sqrt{n_d n_c}) \cdot \exp\left(-\frac{\varepsilon_c - \varepsilon_d}{2kT}\right)$$

Hence the limiting cases for the electrical conductivity of an extrinsic semiconductor are obtained using equation (7.71) in conjunction with the above limiting expressions for the electron concentration in the conduction band. Let us consider these limiting cases. In the case of the high-temperature limit where $T > \theta_D$ we have

$$\sigma = en_d\mu_n$$

ignoring intrinsicity. Using equation (7.76) this is for electron–phonon scattering

$$\sigma \propto n_d T^{-\frac{3}{2}} \quad (7.85)$$

i.e.

$$\ln \sigma \propto -\tfrac{3}{2} \ln T \quad (7.86)$$

However, for the case of low temperatures, if electron-ionized impurity atom scattering is dominant then from equations (7.65) and (7.78) we have

$$\sigma \propto T^{\frac{3}{2}}(\sqrt{n_d n_c}) \exp\left(-\frac{\varepsilon_c - \varepsilon_d}{2kT}\right)$$

i.e.

$$\sigma \propto \sqrt{n_d} T^{\frac{9}{4}} \exp\left(-\frac{\varepsilon_c - \varepsilon_d}{2kT}\right)$$

Once again the exponential term dominates the $T^{\frac{9}{4}}$ term and we find that, effectively

$$\sigma \propto \exp\left(-\frac{\varepsilon_c - \varepsilon_d}{2kT}\right) \quad (7.87)$$

i.e.

$$\ln \sigma \propto \left(-\frac{\varepsilon_c - \varepsilon_d}{2kT}\right) \quad (7.88)$$

Figure 7.10 gives a schematic picture of the temperature variation of the electrical conductivity of an extrinsic semiconductor. It divides into three regions in the same manner as the carrier concentration versus temperature discussed in Section 7.6.

Region 1. The intrinsic range.

Here all extrinsic effects are swamped out as we have discussed in Section 7.6. The electrical conductivity is given by equation (7.82).

Region 2. The saturation range.

Here intrinsic effects are not dominant, all impurity atoms are ionized so that the electron concentration in the conduction band is a maximum (constant) and the electrical conductivity is given by equation (7.85).

The Quasifree Thermally Perfect Electron Gas

Region 3. The low-temperature range.

Here, intrinsic effects are effectively zero and the impurity atoms are only partially ionized. In this case the electrical conductivity is given by equation (7.87).

Table 7.4 gives values of the room-temperature electrical conductivity of some common semiconductors.

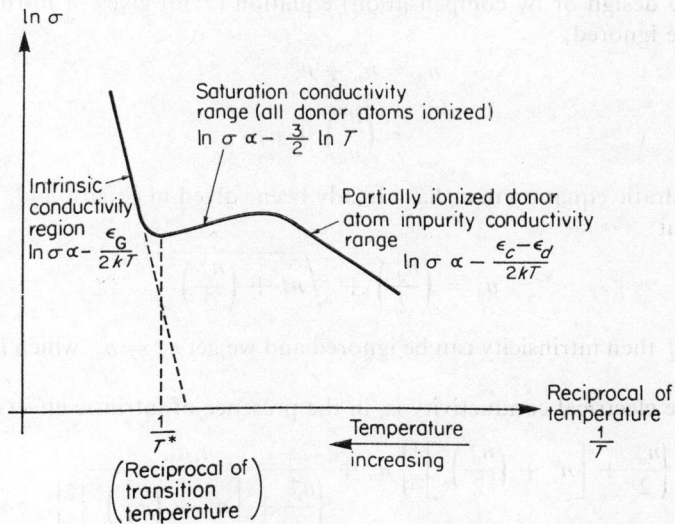

Figure 7.10. Schematic representation of the variation of the natural logarithm of the electrical conductivity of an n-type extrinsic semiconductor as a function of reciprocal temperature. The three distinct conductivity regions shown correspond to the three regions shown in Figure 7.8.

Table 7.4. Values of the electrical conductivity of a selection of semiconductors at room temperature

Material	Electrical conductivity $(\Omega \, m)^{-1}$
Si	$4\cdot 35 \times 10^{-4}$
Ge	$2\cdot 13$
GaSb	10^4
InSb	6×10^4
GaAs	5×10^3
PbTe	50

It will be clear from the equations developed so far that the electrical conductivity of an extrinsic semiconductor must also vary with the level of doping as well as the temperature. Equation (7.71) when combined with equation (7.35) gives

$$\sigma = e\left[n_n\mu_n + \left(\frac{n_i^2}{n_n}\right) \cdot \mu_p\right] \tag{7.89}$$

For an n-type semiconductor with no acceptor impurity atoms (either by deliberate design or by compensation) equation (7.36) gives, if intrinsic effects cannot be ignored,

$$n_n = n_p + n_d^+$$

$$= \left(\frac{n_i^2}{n_n}\right) + n_d^+$$

This quadratic equation in n_n has already been solved in equation (7.52) and we found that

$$n_n = \left(\frac{n_d^+}{2}\right) + \sqrt{n_i^2 + \left(\frac{n_d^+}{2}\right)^2}$$

If $n_i \ll n_d^+$ then intrinsicity can be ignored and we get $n_n = n_d^+$ which is equation (7.63).

Hence the electrical conductivity is, in the presence of intrinsic effects,

$$\frac{\sigma}{e} = \left\{\frac{n_d^+}{2} + \left[n_i^2 + \left(\frac{n_d^+}{2}\right)^2\right]^{\frac{1}{2}}\right\}\mu_n + \frac{n_i^2 \mu_p}{\left\{\frac{n_d^+}{2} + \left[n_i^2 + \left(\frac{n_d^+}{2}\right)^2\right]^{\frac{1}{2}}\right\}}$$

$$= \left\{\frac{n_d^+}{2} + \left[n_i^2 + \left(\frac{n_d^+}{2}\right)^2\right]^{\frac{1}{2}}\right\}\mu_n + \left\{-\frac{n_d^+}{2} + \left[n_i^2 + \left(\frac{n_d^+}{2}\right)^2\right]^{\frac{1}{2}}\right\}\mu_p \tag{7.90}$$

This gives the electrical conductivity in terms of n_d^+, T, ε_G, μ_n, μ_p, and it is usual to plot families of curves of σ versus n_d for different ε_G and T values. We shall not give such graphical plots here since they are to be found in the standard text books. Table 7.5 gives values of the electrical conductivity of silicon and germanium at room temperature as a function of the doping level. These figures are only approximate and should be taken as showing no more than the trend of the change in conductivity rather than the exact value. Experimental studies of the electrical conductivity of extrinsic semiconductors show that the electrical characteristics are dependent on how the semiconductor has been prepared and stored and are not exactly reproducible unless care is taken in these matters.

Now that we have determined the electrical conductivity of both intrinsic and extrinsic semiconductors, it is of interest to enquire if the Wiedemann–Franz–Lorenz law of Section 6.9 can be applied to these semiconductors and in particular to see if equation (6.72) is obeyed, viz.

The Quasifree Thermally Perfect Electron Gas

$$\frac{\rho_T(\lambda_e)_T}{T} = \frac{\pi^2}{3}\left(\frac{k}{e}\right)^2 = L = 2\cdot 45 \times 10^{-8} \text{ W }\Omega \text{ deg C}^{-2}$$

For those semiconductors where the classical approximation of Section 7.3 is valid it is found that the left-hand side of the above equation is constant but the numerical value of the constant is not equal to the Lorenz number L—it is

Table 7.5 The electrical conductivity of n-type silicon and germanium in $(\Omega \text{ m})^{-1}$ at room temperature as a function of the doping concentration.

Material	Intrinsic	\multicolumn{6}{c}{Value of doping concentration in m^{-3}}					
		10^{16}	10^{18}	10^{20}	10^{22}	10^{24}	10^{25}
n-type Si	$4\cdot 4 \times 10^{-4}$	5×10^{-4}	2×10^{-2}	2	$1\cdot 7 \times 10^2$	10^4	$1\cdot 7 \times 10^4$
n-type Ge	$2\cdot 1$	$2\cdot 1$	$2\cdot 1$	5	5×10^2	5×10^4	$1\cdot 3 \times 10^5$

found to be larger. This discrepancy is of considerable academic interest because, unlike metals the lattice contribution to the total thermal conductivity is significant and we can therefore study lattice dynamics; in addition, unlike insulators semiconductors can be made to a very high state of purity and relatively free of lattice defects, thereby allowing phonon-scattering mechanisms to be studied. However, these aspects are of little significance to this textbook on thermodynamics. When the classical approximation is valid we have shown in equation (7.30) that the intrinsic carrier concentration is 'low' so that the resistivity is 'quite large' and therefore the thermal conductivity is 'small', i.e. the valence-electron gas thermal conductivity is small and is in fact completely dominated by the lattice thermal conductivity. It is on the basis of this argument that we can divide semiconductors into roughly two classes when we consider their thermal conductivities. Firstly there are the semiconductors of 'low' carrier concentration such as silicon and germanium. As we have just remarked for such semiconductors the valence-electron gas thermal conductivity is small so that from equation (6.64) we can sensibly ignore it and write,

$$\lambda_T \sim \lambda_L$$

which is the criterion that holds for insulators [see equation (6.67)]. For example experiment shows that

$$\frac{\lambda_L}{\lambda_e} = 377 \text{ for Si}$$

$$= 105 \text{ for Ge.}$$

Consequently we would expect the thermal conductivity versus temperature behaviour to be similar to that of Figure 6.4 and this is found to be the case. The effect of doping is two-fold.

1. It increases the carrier concentration considerably above the intrinsic value and therefore increases both the electrical conductivity and the valence-electron gas thermal conductivity.
2. It reduces the thermal conductivity of the lattice by causing phonon-impurity atom scattering analogously to the discussion in Section 6.7.

In the limit there are also the heavily doped semiconductors where the carrier concentration is 'high'. In this case the electrical resistivity is 'low' so that the valence-electron gas thermal conductivity makes a non-negligible contribution to the total thermal conductivity. For example, at room temperature in the case of PbTe we have

$$\frac{\lambda_L}{\lambda_e} = 7$$

Nevertheless, the lattice thermal conductivity still tends to contribute more to the total thermal conductivity even in this heavily doped case.

Table 7.6 gives some values of the thermal conductivity of typical semiconductors and these should be compared with Tables 6.2 and 6.3.

Table 7.6 Values of the total thermal conductivity of a selection of semiconductors at room temperature.

Material	Debye temperature °K	Total thermal conductivity λ_T W/m deg C at 300°K
Si	640	137
Ge	370	54
GaSb	266	33
InSb	200	17
GaAs	344	44

The reader may wonder why we should wish to know about the thermal conductivity of semiconductors. In fact the thermal conductivity is very important because, as we saw at the end of Section 7.6, the operating temperature of semiconductor devices must be carefully watched to avoid either mechanically damaging the device, melting the metallic solders or changing the carrier concentrations. In turn, the operating temperature will be dictated by the

The Quasifree Thermally Perfect Electron Gas

power being dissipated in the device; as this power increases, so the junction temperature will rise. The best way to dissipate this joule heat and therefore keep the semiconductor device within the allowed range of operating temperature is by thermal conduction. In essentially all transistors the flux of thermal energy (heat flow) from the junction is proportional to the temperature difference between the junction and the transistor case. This follows from equation (6.3), viz.

$$\text{Flux of thermal energy} = -\lambda \frac{dT}{dx}$$

Consider the flow of thermal energy from the junction to the case; any small-volume element of semiconductor in the path of this flow of thermal energy can be considered a thermodynamic system to which the First Law of Thermodynamics is applicable. In integrated form this is given by equation (1.2) and we can place it on a time basis to give

$$\dot{Q} - \dot{W} = \Delta \dot{E}$$

In a steady-state condition where the heat and work transfers to and from the system exactly balance, the thermodynamic energy of the system will not change, i.e. $\Delta \dot{E} = 0$.

Also, from equation (6.3)

$$\dot{Q} = -\lambda A \frac{dT}{dx}$$

where A = surface area normal to the flow of thermal energy. In addition

$$\dot{W} = -P$$

where P is the electrical power supplied to the semiconductor and the minus sign allows for the fact that electrical power is supplied *to* the system. Hence using the First Law and equation (6.3) we get for the steady-state condition,

$$-\lambda A \frac{dT}{dx} - (-P) = 0$$

i.e.

$$\frac{dT}{dx} = \frac{P}{\lambda A}$$

This is the differential equation that gives the temperature gradient $\frac{dT}{dx}$ between the junction and the semiconductor case. Although Chapters 6 and 7 show that metals and semiconductors in general have temperature-dependent thermal conductivities, because of the fact that the operating temperature range of the semiconductor is not large in many applications we can, to a first approximation, treat the thermal conductivity as a constant independent of temperature over the operating temperature range. If we are allowed to make this approximation, then the temperature gradient is a constant and we can integrate the above equation to get

$$\int_{T_c}^{T_J} dT = \frac{P}{\lambda A} \int_0^L dx$$

where T_J and T_c are the respective temperatures of the junction and the case and L is the distance from the junction to the case, i.e.

$$T_J - T_c = \frac{P}{\lambda A} \cdot L$$

This tells us that the temperature rise at the junction is proportional to the power supplied; in addition the proportionality factor is $\frac{1}{K}$ where

$$\frac{1}{K} = \frac{L}{\lambda A}$$

K is known as the thermal conductance (*not* thermal conductivity) of the semiconductor. Just as the reciprocal of thermal conductivity λ was thermal resistivity W so the reciprocal of thermal conductance K is thermal resistance R_{th}.
∴

$$T_J - T_c = R_{th} \cdot P$$

Clearly R_{th} gives the temperature difference between two points in the semiconductor when one watt (1 J/s) is being dissipated.

In practise a calculation of the temperature rise of a semiconductor device when power is being supplied can be fairly complicated because the thermal conduction must be through not only the semiconductor material but also through the thin layers of solder, through the metal case into the heat sink which is merely a device constructed to maximize the 'heat' that it rejects to the surrounding atmosphere. The thermal resistance of this complex collection of electrical and thermal conductors can be calculated using the electrical circuit analogue of thermal circuits which is properly a branch of heat transfer discussed in standard textbooks on that topic.

Knowing the dimensions of a semiconductor we can calculate values of the thermal resistance. Table 6.2 shows that the thermal conductivity of metals lie within the range 100–300 W/m deg C whilst Table 7.6 shows that the thermal conductivities of semiconductors are somewhat lower. Values of the thermal

The Quasifree Thermally Perfect Electron Gas

resistance are usually given by the manufacturer of the semiconductor device and range from 0·2 deg C/W for power transistors with large areas and efficient heat sinks to as high as 1000 deg C/W for narrow-base, low-power alloyed transistors.

7.9 Diffusion processes in a semiconductor

In Chapters 6 and 7 we have discussed the problem of electrical current flow in a metal and in a semiconductor due to an externally applied electric field. However, it is also possible to produce an electrical current flow in the *absence* of an externally applied electric field if we arrange that the carrier concentration in the conductor at any one point exceeds the carrier concentration at some other point in the same conductor.

Equation (6.4) stated Fick's law of diffusion,

$$\text{Flux of carriers} = - D_\alpha \frac{dn_\alpha}{dx}$$

Here the flux of carriers is the number of carriers crossing unit area of the conductor per unit time. D_α is the transport coefficient known as the diffusion coefficient and has units of $m^2 s^{-1}$. $\frac{dn_\alpha}{dx}$ is the spatial gradient of the carrier concentration in the conductor. The minus sign in equation (6.4) makes the right-hand side positive (the left-hand side is positive) because n_α decreases as x increases, so that $\frac{dn_\alpha}{dx}$ is negative. If we multiply the carrier flux by the electron charge then we obtain the carrier-current density due to diffusion

Carrier-current density
in the x direction due to = flux of carriers × charge on each carrier
diffusion

$$j_\alpha = - D_\alpha \frac{dn_\alpha}{dx} \cdot Z_\alpha e$$

$$= - D_\alpha Z_\alpha e \frac{dn_\alpha}{dx} \qquad (7.91)$$

where $Z_\alpha = -1$ for electrons and $+1$ for holes. Hence

Electron current density due to diffusion

$$= j_n = D_n e \frac{dn_n}{dx} \qquad (7.92a)$$

and

Hole-current density due to diffusion

$$= j_p = -D_p \cdot e \frac{dn_p}{dx} \quad (7.92b)$$

If we so wish we can [analogously to equation (6.7), which gave the carrier drift velocity in an externally applied electric field] introduce the carrier diffusion velocity in the internal concentration gradient thus

$$j_\alpha = n_\alpha Z_\alpha \cdot e \cdot (v_\alpha)_{\text{diff}} \quad (7.93)$$

whence using equation (7.91) we get

$$(v_\alpha)_{\text{diff}} = -D_\alpha \cdot \frac{1}{n_\alpha} \cdot \frac{dn_\alpha}{dx} \quad (7.94)$$

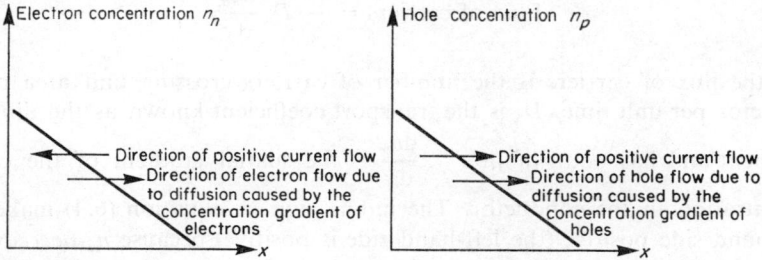

Figure 7.11. Sign conventions for charge-carrier movement and positive electrical current flow for the diffusion of electrons and holes in a semiconductor due to the presence of an internal carrier concentration gradient.

The difference in signs between equations (7.92a) and (7.92b) is related to the difference between the direction of carrier flow in the concentration gradient and the sign convention for positive current flow. The situation is summarized in Figure 7.11.

Clearly from equations (7.92a) and (7.92b) the carrier current density due to diffusion will only exist if there is a concentration gradient in the semiconductor. Such a concentration gradient can arise in practise either due to having non-homogeneous specimens of semiconductor material or actually creating an excess of charge carriers in some region of the semiconductor by either carrier injection, irradiation of the semiconductor with light or by heating. Numerical values of D_n at room temperature are given in Table 7.7 for a selection of semiconductors.

Perhaps the best known application of the semiconductor is the transistor which is made by sandwiching a thin layer of p-type material (base) between

The Quasifree Thermally Perfect Electron Gas

two slabs of n-type material (emitter and collector). Clearly the operation of the device is governed by the behaviour of the two p–n junctions. We shall not analyse the p–n junction since this is a problem in solid-state theory that is given in all the standard text books. However, thermodynamics does make a profound contribution to the analysis because it states unambigously what the behaviour of the Fermi levels on both sides of the junction *must* be.

Table 7.7 Numerical values of the diffusion coefficient of a selection of n-type extrinsic semiconductors at room temperature.

Material	Diffusion coefficient D_n m² s⁻¹
Si	$3 \cdot 1 \times 10^{-3}$
Ge	$9 \cdot 3 \times 10^{-3}$
GaSb	$1 \cdot 0 \times 10^{-2}$
InSb	$9 \cdot 3 \times 10^{-3}$
GaAs	$7 \cdot 0 \times 10^{-3}$
PbTe	$5 \cdot 2 \times 10^{-3}$

In equation (2.88) we showed that the Lagrange undetermined multiplier α was constant across the junction of two thermodynamic systems provided that the two systems were in thermodynamic equilibrium. Equation (4.1) expressed α in terms of the Fermi energy so that we can say that

$$\alpha_n = \alpha_p$$

$$\frac{(\varepsilon_F)_n}{kT_n} = \frac{(\varepsilon_F)_p}{kT_p}$$

Since the temperature of both systems is the same (thermal equilibrium being a necessary part of thermodynamic equilibrium (see Section 1.4)) then

$$(\varepsilon_F)_n = (\varepsilon_F)_p \tag{7.95}$$

i.e. the Fermi energies are equal and the *Fermi levels align*. Of course the conduction and valence band edges do *not* align and it is just this non-alignment that produces the built-in or pre-bias steady potential (contact potential) that is so important for p–n junction working.

Problems for Chapter 7

1. A piece of very pure germanium is intrinsic at 300°K. Find:
   a. The intrinsic electron and hole concentrations.
   b. The fraction of current carried by the electrons.
   c. The electrical conductivity.

d. The current density flowing in the germanium if an external field of 10^4 V m is applied to the germanium crystal.
e. The electron and hole drift velocities in this applied electric field.
f. The root mean square thermal speed of the electrons and holes.
g. The electron and hole collisional relaxation times.

You may assume that the energy gap is 0·65 eV, that the electron and hole mobilities are 0·36 and 0·17 m²/V s respectively and that both carrier effective masses are equal to the isolated electron mass.

Answers: 8.77×10^{19} m^{-3}; 68%; 7·44 $(\Omega$ m$)^{-1}$; 7.44×10^4 Am^{-2}; -0.36×10^4 ms^{-1}, 1.7×10^3 ms^{-1}; 1.17×10^5 ms^{-1}; 2.05×10^{-12} s; 9.67×10^{-13} s.

2. A piece of silicon of atomic weight 28 kg (kg mole)$^{-1}$ and density 2.33×10^3 kg m^{-3} at 300°K contains 126 atoms of arsenic in 10^8 host atoms of silicon such that the arsenic is 90% ionized. If the experimentally measured values of the energy gap and intrinsic carrier concentration are 1·15 eV and 1.4×10^{16} m^{-3}, find:
a. The electron density in the conduction band.
b. The hole density in the valence band.
c. The electrical conductivity of the semiconductor.
d. The position of the Fermi level with respect to the conduction-band edge.
e. If the silicon had contained 9.3×10^{22} ionized arsenic donors per cubic metre and 3.6×10^{22} ionized gallium acceptors per cubic metre, how would this have affected the electrical conductivity?

You may assume that the electron and hole mobilities are 0·16 and 0·04 m²/V s and that their effective masses are equal.

Answers: 5.7×10^{22} m^{-3}; 3.4×10^9 m^{-3}; 1.5×10^3 $(\Omega$ m$)^{-1}$; 0·21 eV below the band edge; no change in electrical conductivity.

CHAPTER EIGHT

Information Theory and Thermodynamics

8.1 Information theory and thermodynamics

The preceding chapters have been devoted to the application of statistical thermodynamics to metals and semiconductors. However, the reader should not conclude that thermodynamics is solely limited to considering the thermodynamic and electrical properties of these materials.

In this chapter we shall break away entirely from solid-state theory and briefly introduce an application of thermodynamics of interest to the electrical engineer and physicist and yet one which is very controversial. It must be stressed that this chapter is in no way related to the topics discussed in Chapters 3–7.

The basic concept underlying information theory is that information can be defined and treated as a physical quantity that can be generated and measured (in an information source), transmitted or communicated via a physical medium (cable, radio link, mercury line, magnetic tape, etc.) and received (by a receiver which may be an instrument or a human being). The actual method of communication will require a finite set of symbols that is appropriate to the type of communication channel being used, for example, the morse code of dots and dashes and spaces is one set of 'symbols' suitable for wireless telegraphy, whilst the alphabet is another set of symbols suitable for a teacher at the blackboard. Fishes possibly communicate with each other using a collection of sounds of different frequency and intensity. Whatever the symbols may be, they are completely independent of each other and an arbitrary succession of such symbols is called a message. Of course, such a message may or may not be intelligible to the receiver. For example, a human being may not understand the 'language' of fish or the writings of a monkey trained to draw the symbols of the alphabet on a piece of paper!

Clearly then, for a given set of symbols many messages, intelligible or otherwise to a receiver, may be sent. Information theory concerns itself with any one particular message that can be chosen from this finite set of possible messages when this message is transmitted through the communication channel and received by a receiver. Since there is a choice of messages from the finite set we can say that each alternative has a probability of occurring. For example, consider the situation where some event has happened and it is necessary to transmit a message using an information source via a communication channel to a receiver such that the receiver is to be informed about the event having taken place. Clearly the important thing about the message is the order and arrangement of the symbols transmitted. Before the message is received, the receiver is in doubt or uncertain about whether the event has happened or not, and if so, what its outcome is. After the message is received the receiver should ideally be perfectly certain, that is, have no doubt or uncertainty at all about the outcome of the event. If this is so, then clearly the transmission of the message has removed uncertainty about the event having occurred and about its outcome. Let us define

Initial uncertainty of the receiver about the outcome of the event before the message is sent

$$= -K \log p_i \tag{8.1}$$

Here K is a constant whilst p_i is the probability of the outcome of the event about which the message is being sent before receipt of the message. Since p_i is always less than unity, its logarithm (whatever the base) is negative and the initial uncertainty always positive. Now suppose the message is sent such that after its receipt by the receiver, the receiver is still not completely certain about the outcome of the event. Then we have

Final uncertainty of the receiver about the outcome of the event after the message is sent

$$= -K \log p_f$$

where p_f is the new probability of the outcome of the event after receipt of the message. We define

Information in the message $= I$

$\qquad = $ *decrease* in the uncertainty of the receiver

$\qquad = -$ (final uncertainty $-$ initial uncertainty)

$\qquad = -[-K \log p_f - (-K \log p_i)]$

$$= -K \log \frac{p_i}{p_f} \tag{8.2}$$

The reader may be puzzled as to why after the message has been sent, the receiver may still be uncertain as to the outcome of the event. For an ideal loss-less, noise-free communication channel, the receiver would be in no doubt, i.e. he would be completely certain about the outcome of the event. Hence the

Information Theory and Thermodynamics

probability of the event after receipt of the message is unity ($p_f = 1$) because there are no alternatives. In this case

$$I = -K \log \frac{p_i}{1}$$

$$= -K \log p_i \qquad (8.3)$$

However, in a real-life communication channel, noise or random interference may make it impossible for the receiver to be completely certain about the outcome of the event even after receipt of the message. For example, he might be receiving a sequence of abbreviated words, some of which may be obscure or even meaningless if the channel is noisy.

In the above definition we have defined uncertainty only for the case where there are Ω *equally likely* outcomes for the message such that $p = \frac{1}{\Omega}$ and there is no particular outcome that is more favoured than any other. In this case equation (8.3) becomes

$$I = -K \log p_i$$

$$= -K \log \left(\frac{1}{\Omega}\right)$$

$$= K \ln \Omega \qquad (8.4)$$

Such is not usually the case, however. Consider, for example, a written sentence in the English language. The problem is to calculate the amount of information in the sentence bearing in mind that the sentence is formed from only 27 symbols (26 letters and a blank space). If these 27 symbols were equally probable, that is equally likely to appear in words, then the number of distinctly different messages (regardless of their intelligibility or value) that can be composed using a total of N symbols drawn from a pool of 27 symbols is 27^N i.e.

$$\Omega = 27^N$$

or

$$p_i = \frac{1}{27^N}$$

whence

$$I = KN \log 27 \qquad (8.5)$$

and average information per symbol

$$= \frac{I}{N} = K \log 27 \qquad (8.6)$$

If we use natural logarithms then this has the numerical value 3·3 K. However we know that the different letters of the alphabet do *not* occur equally probably. The blank space occurs most frequently, followed by the letters E, T, O, A, ... and by sentence analysis we can actually determine the probability, the numerical value of the relative frequency of occurrence of each symbol. Hence the problem is to decide the information content of N symbols with different probabilities of occurrence. This is merely the problem of finding the number of permutations of N things of which N_1 are alike (for instance, the blank spaces), N_2 are alike (such as the letter E), etc. such that $\Sigma N_i = N$. This number is $\dfrac{N!}{\Pi N_i!}$ and this represents the total number of messages that can be sent so that its reciprocal is the probability of any one message, thus the information sent in any one message is

$$I = -K \log \frac{\Pi N_i!}{N}$$

$$= -K\Sigma \ln N_i! + K \ln N!$$

Now by Stirling's theorem (see discussion on page 44)

$$\log N! = N \ln N - N$$

provided $N > 10$, say

$\therefore$

$$I = -K[\Sigma N_i \log N_i - \Sigma N_i - N \log N + N]$$

$$= -K\Sigma N_i \log \frac{N_i}{N}$$

Hence

Average information per symbol

$$= \frac{I}{N}$$

$$= -K\Sigma \frac{N_i}{N} \log \frac{N_i}{N} \tag{8.7}$$

But $\dfrac{N_i}{N}$ = probability that the receiver would assign to a symbol identified by subscript i.

$$= p_i$$

whence

$$\text{Average information per symbol} = -K\Sigma p_i \log p_i \tag{8.8}$$

For the alphabet problem this can be shown to have the numerical value of 2·79 K which should be compared with 3·3 K derived earlier for symbols of equal frequency. We therefore see that in the case of equal frequency of occurrence the amount of information per symbol is more than the unequal frequency of occurrence case, thus the *more* we know about the frequency of occurrence of symbols, the *less* information can we communicate using these symbols. This perhaps unexpected result arises from the fact that the more we know about the outcome of an event the less information can we receive about its actual outcome.

Equation (8.8) is known as Shannon's formula and we see that the average information per symbol depends upon the probabilities of all of the symbols. In fact, this result is quite general in information theory, as the average amount of information per symbol is given by equation (8.8) and the more that is known the less the information that can be communicated.

The reader may quite rightly wonder what all of this has got to do with thermodynamics. A full discussion of this point revolves around equation (8.8) and the mathematical expression for entropy that is given by the statistical thermodynamics of the canonical ensemble. We shall not give this discussion nor even attempt to define the meaning of 'canonical ensemble'. What we can do, however, is to consider the statistical thermodynamics already introduced in Chapter 2.

8.2 The meaning of Entropy; the Third Law of Thermodynamics

One of the links between the macroscopic thermodynamic properties of a system and its microscopic aspects is equation (2.49) viz.

$$S = k \ln \Omega + S_0$$

where S_0 is a constant and where Ω has been defined as the total number of microstates for all macrostates of a thermodynamic system consisting of a thermally perfect gas of N particles confined to a volume V and having a total energy E. Suppose we cool down this thermally perfect gas and thereby reduce the total energy E. Clearly since the particles possess only translational energy (non-interacting particles with zero potential energy of interaction) and possibly internal energy due to their structure (see Chapter 3), as the gas is cooled the particles will seek translational energy states of lower and lower energy. However, since the particles do not interact, they will not form a condensed phase (liquid or solid) and the system will remain as a gas. As the temperature falls towards absolute zero all particles will end up in the lowest translational energy states—provided that there are no restrictions on the occupancy of such translational energy states.

In Chapter 4 where we discussed the F.D. valence-electron gas in a metal, we showed that even at the absolute zero of temperature the particles were spread out over a vast number of translational energy levels ranging from translational energy zero to the Fermi energy, thus all levels were filled to their Pauli quota. In addition, for all translational energy levels less than the Fermi energy $N_n = g_n$ since this *is* the Pauli quota. If we combine equations (2.23) and (2.27) with the result that $N_n = g_n$ we obtain

$$\Omega_{FD} = \prod_n \frac{1!}{0!}$$

$$= 1 \text{ since } 0! = 1$$

whence from equation (2.49)

$$S_{FD} = k \ln 1 + S_0$$
$$= S_0 \text{ since } \ln 1 = 0$$

What is the meaning of $\Omega = 1$? What this means is that there is only *one* microstate and only *one* macrostate and there is no other way of distributing the fermions over the allowed accessible translational energy levels.

Now consider the thermally perfect BE gas where there is no limit to the number of particles that we can put into any translational energy state; let us cool such a gas down to the absolute zero. In this case *all* the particles will crowd into the translational energy level of lowest energy viz. the level with translational energy ε_0 and degeneracy g_0. If we put this result into equations (2.21) and (2.27) we obtain

$$\Omega_{BE} = \frac{(g_0 + N - 1)!}{(g_0 - 1)!N!}$$

Table 2.1 shows that the ground-level degeneracy is unity so that

$$\Omega_{BE} = 1 \text{ and } S_{BE} = S_0$$

Once again we see that there is only *one* microstate and only *one* macrostate available for the particles.

The Third Law of Thermodynamics states that a thermodynamic system which is in true thermodynamic equilibrium and in its lowest energy condition has both zero absolute temperature and zero entropy, i.e. $T = 0$ and $S = 0$ so that $S_0 = 0$. This has been a controversial law in the past because many thermodynamic systems made up of interacting particles (that is, *not* thermally perfect gases) do not remain in total thermodynamic equilibrium when cooled through the liquid and solid phases. Nevertheless we shall accept the law as stated. The equation

$$(\Omega_{BE/FD})_{T=0°K} = 1$$

Information Theory and Thermodynamics

is interpreted as meaning that at the absolute zero of temperature the gas is in a state of 'complete order or zero randomness or zero uncertainty'. This interpretation is purely on the basis that there is only one microstate and only one macrostate. However, for all finite non-zero temperatures the energy of the system rises above the minimum energy such that essentially all of the particles increase their energy from the value that they had at the absolute zero and consequently spread themselves out over more and more translational energy levels because these levels have now become accessible to the particles because of the increased energy of the particles. This means that the number of microstates increases very, very rapidly as the temperature is raised, because from equations (2.49) and (2.52) and the fact that S_0 equals zero we can write

$$\Omega_{BE/FD} = \exp \frac{E}{kT} \exp Z_{BE/FD}$$

i.e. the number of microstates $\Omega_{BE/FD}$ is exponentially dependent on the energy.

As the particles spread themselves over an increasing number of translational energy levels, these becoming accessible as the temperature of the system is raised, so we are less sure as to where they are, thus our 'complete order or zero randomness or zero uncertainty' gives rise to 'disorder or randomness or uncertainty' purely because as Ω increases so we become less and less able to say in which particular translational energy levels the particles are in due to both their increasing accessibility and increasing degeneracy. In addition, since our equilibrium analysis of the most probable macrostate in Section 2.10 was based on the supposition that the gas would seek that macrostate with the *greatest* number of associated microstates, we see that the equilibrium condition is the condition of maximum 'disorder or randomness or uncertainty'.

Thus we can equate what is microscopically the tendency of the system to move towards the most probable macrostate to what is macroscopically the entropy increase of the system and it is just this equivalence that is used to give a 'physical picture' of entropy. However, we must be careful to appreciate what we are doing. Macroscopically speaking, *entropy is a purely mathematical function with no macroscopic interpretation.* Undoubtedly the change in the entropy of a system given by equation (1.7) viz.

$$dS = \frac{dQ_R}{T}$$

in terms of the reversible heat transfer to the system is a well-defined quantity which can be *measured* between any two equilibrium thermodynamic states; however, equation (1.7) gives us no insight into what S represents. If we want such insight then we have to define entropy statistically by saying that entropy is proportional to the number of microstates accessible to the system—which is exactly equation (2.49).

The reason that entropy has no simple macroscopic interpretation is related to the fact that it does not obey the conservation criterion, that it is not conserved but is constantly being created by spontaneous processes. We think that we have no difficulty in interpreting the conserved quantities of physical science such as mass, momentum and energy and we go to great lengths to preserve the idea of conservation from the macroscopic level right down to the microscopic level. However, entropy stubbornly refuses to be conserved, and just continually increases. Its continual increase presents a mental stumbling block to our accepting entropy as just another thermodynamic property like energy or enthalpy.

8.3 Entropy and Information

We can now return to our discussion of information theory by noting that there is a link between the two due to the obvious similarity of equations (8.4) and (2.49) with $S_0 = 0$ viz.

$$I = K \ln \Omega$$
$$S = k \ln \Omega$$

In the former equation Ω is the number of equally likely outcomes for the message, whilst in the latter Ω is the number of equally likely (equal a priori) microstates describing the system. In addition we have discussed an increase of information as a loss of uncertainty. We can invert this argument and say that a loss of information is an increase of uncertainty. However, we have also discussed the concept of the increase of entropy of a system as being an increase of 'disorder of randomness or uncertainty'. At first sight there appears to be a great deal of similarity between I and S.

A more detailed discussion of this similarity has led some writers to assume that perhaps information theory is a very basic scientific discipline, perhaps even more basic and primitive than thermodynamics itself. Developing this viewpoint has led writers to make a one-to-one identification of the two disciplines and thereby obtain such equations as

$$\text{Loss of information} = \text{increase in uncertainty}$$
$$= \text{increase of entropy} \qquad (8.9)$$

They have used this identification in a more sophisticated form to deduce all the formulae of statistical thermodynamics and therefrom to derive all of the laws of macroscopic, or phenomenological thermodynamics.

It is appropriate to end this book on a controversial note if for no other reason than to show that the subject of thermodynamics is still far from complete and its true role in the world far from being completely appreciated. As a conservative professional thermodynamicist, the writer is not in sympathy with those who equate entropy and information content. He feels that it would be better to rewrite the latter phrase of equation (8.9) as 'increase of communication entropy'. Perhaps all we should really do is to draw an *analogy* rather than form an equality between information I and *statistical* entropy $k \ln \Omega$, and then continue the analogy by relating *statistical* entropy to *thermodynamic* entropy. This type of reasoning would be applicable to any physical system which has a finite number of accessible states weighted with different probabilities and not just the quantum levels and microstates of a thermodynamic system. Such physical systems are abundant in our world and we could therefore have some form of analogy between statistical entropy and variables in biological systems, psychological systems, economic systems and syntactic systems.

Finally, the writer does not agree with the view that information theory is more primitive and basic than thermodynamics. Both information theory and statistical thermodynamics are situations requiring statistical methods which, of course, as we have just seen, have a lot in common. However, the writer has an increasingly profound respect for macroscropic (phenomenological) thermodynamics as opposed to statistical thermodynamics as he learns and understands more about it, and feels that it may occupy a rather special place in the 'order of things'. Perhaps Einstein should be allowed the last word; he wrote 'A theory is the more impressive the greater the simplicity of its premises, the more different kinds of things it relates and the more extended its area of applicability; hence the deep impression that macroscopic thermodynamics made upon me. It is the only physical theory of universal content concerning which I am convinced that, within the framework of applicability of its basic concepts, will never be overthrown'.

APPENDIX ONE

Some Physical Constants in SI Units

Values are rounded off to two decimal places.

Planck's constant h	$6 \cdot 63 \times 10^{-34}$ J.s
Elementary charge e	$1 \cdot 60 \times 10^{-19}$ C
Boltzmann's constant $k = \dfrac{R}{N_{AV}}$	$1 \cdot 38 \times 10^{-23}$ J°K^{-1}
Velocity of light c	$3 \cdot 00 \times 10^{8}$ m/s
Mass of electron m_e	$9 \cdot 11 \times 10^{-31}$ kg
Mass of hydrogen atom m_H	$1 \cdot 67 \times 10^{-27}$ kg
Universal gas constant R	$8 \cdot 31 \times 10^{3}$ J (kg mole °K)$^{-1}$ or J (kilomol °K)$^{-1}$
Avogadro number N_{AV}	$6 \cdot 02 \times 10^{26}$ (kg mole)$^{-1}$
Specific volume of a thermally perfect gas at STP viz. 0°C and 1 atmosphere	$22 \cdot 41$ m^3 (kg mole)$^{-1}$
Atmospheric pressure p_{AT}	$1 \cdot 01 \times 10^{5}$ N/m^2

In the book we have used the energy-conversion factor

$$1 \text{ eV} = 1 \cdot 60 \times 10^{-19} \text{ J}$$

APPENDIX TWO

The Mole, Atomic Weight and Molar Volume

Consider a system comprising an arbitary amount of matter. Let

M = total mass of system in kg

N = total number of atoms in the system

N_e = total number of valence electrons in the system

m = mass of one atom in kg

A = atomic weight of the system which by definition is the mass of one mole of the system in the appropriate mass units, i.e. in the present case, kilograms per kg mole

V = volume of the system in m^3

ρ_D = density of the system in kg/m^3

N_{AV} = Avogadro number

A kilogram mole i.e. a kg mole (or, as it should be correctly known, kilomol) of substance is the chemical unit for the amount of matter present. It is not a unit of mass but a unit of measure in that it refers to the number of atoms or molecules present. It is defined as the amount of matter (by accumulation and not by mass or weight) that has a mass in kilograms equal to the atomic or molecular weight of the substance, i.e.

1 kg mole or (kilomol) of substance = A kg of that substance

1 kg mole contains the Avogadro number of atoms or molecules viz. $6 \cdot 02 \times 10^{26}$.

By definition,

$$\begin{aligned} A &= \text{mass in kg of 1 kg mole of system} \\ &= \text{number of atoms in one kg mole} \times \text{mass of an atom} \\ &= N_{\text{AV}} \times m \end{aligned} \quad (A1)$$

Also,

Total mass of system = number of atoms in system × mass of one atom

$$M = Nm \tag{A2}$$

Eliminate m from (A1) and (A2) to get

$$\frac{A}{N_{AV}} = \frac{M}{N}$$

i.e.

$$N = \frac{M}{A} \cdot N_{AV} \tag{A3}$$

Also, if we divide (A3) by V we get

$$\frac{N}{V} = N_{AV} \cdot \frac{M}{V} \cdot \frac{1}{A}$$

i.e.

$$\frac{N}{V} = \frac{N_{AV} \cdot \rho_D}{A} \tag{A4}$$

Finally,

Valency of an atom = average number of valence electrons per atom

Hence

$$N_e = \text{Valency} \times N$$

$$= \text{Valency} \times N_{AV} \times \frac{M}{A} \tag{A5}$$

from (A3).

Also,

Valence-electron concentration or number density

$$= \frac{N_e}{V} = \text{Valency} \times \frac{\rho_D}{A} \times N_{AV} \tag{A6}$$

Example:

The atomic weight of lead is 207·2 kg (kg mole)$^{-1}$, its valency is 4 and its density at 20°C is 11·34 × 10^3 kg/m^3. Find for lead at 20°C,

a. the number of atoms per cubic metre
b. the number of valence electrons per cubic metre (electron density)
c. the molar volume of lead

Appendix 2

(a) From (A4)

Number of atoms per cubic metre

$$= \frac{\rho_D}{A} \cdot N_{AV}$$

$$= \frac{11\cdot 34 \times 10^3 \times 6\cdot 02 \times 10^{26}}{207\cdot 2}$$

$$= 3\cdot 30 \times 10^{28} \text{ atoms per cubic metre}$$

(b) From (A6)

Valence-electron concentration or number density

$$= \text{Valency} \times \text{Number of atoms per cubic metre}$$
$$= 4 \times 3\cdot 30 \times 10^{28}$$
$$= 1\cdot 32 \times 10^{29} \text{ electrons per cubic metre}$$

(c) Molar volume = Volume per kg mole

$$= \frac{\text{Mass of one kg mole}}{\text{Density}}$$

$$= \frac{\text{Atomic weight}}{\text{Density}}$$

$$= \frac{A}{\rho_D}$$

$$= \frac{207\cdot 2}{11\cdot 34 \times 10^3}$$

$$= 1\cdot 83 \times 10^{-2} \text{ m}^{-3}$$

Index

Absolute temperature, *see* Temperature
value of entropy, *see* Entropy
Acceptor atoms, *see* Extrinsic semiconductor
Alkali metals, 80, 92, 99, 193
Anharmonic oscillator, 160, 173, 174, 175, 176
Atomic energy levels, *see* Electron energy states of an atom
Atomic nucleus, 74
Atomic weight, applied to solids, 154, 266, 267
  definition of, 265
Average numbers of phonons per oscillator, 133, 134, 152
  thermal energy, *see* Energy, average thermal
Avogadro hypothesis, 51, 266, 267
  number, 105, 264

Band theory of solids, *see also* Carrier concentrations,
  Bloch potential, 191, 199
  conduction band, 191, 192, 193, 194, 196, 197, 199, 200, 202, 204, 206, 208, 210, 211, 212, 229, 237, 241, 253
  energy gap, 191, 192, 193, 194, 195, 196, 197, 198, 201, 208, 209, 211, 212, 214, 216, 218, 225, 231, 241, 242
  energy gap, table of values, 212
  valence band, 191, 192, 193, 194, 195, 196, 197, 198, 199, 200, 203, 204, 207, 208, 210, 211, 212, 230, 237, 241, 253
Binding energy, *see* Extrinsic semiconductor
Black body, radiation energy distribution, 107, 111, 112, 113, 116, 126

radiation field and radiation oscillator, 106, 107, 111, 112
Bloch theory, *see* Band theory of solids
Bohr orbit, *see* Electron energy states of an atom
Boltzmann distribution law, 60
  transport equation, 138
Boltzmann's constant, numerical value of, 264
Bose–Einstein, distribution function, 49, 55, 118, 119, 133, 134
  phonon gas, *see* Lattice, Debye theory
  photon gas, 113, 126
  statistics, 33, 41, 46, 47, 48, 50, 58, 113, 118, 126
Bosons, 33, 37, 39, 118
Boundary conditions, periodic, 82, 83, 140, 149
  stationary, 16, 17, 82, 83, 140
Boundary scattering, of electrons, 158, 167, 183
  of phonons, 179, 182
Box, particle in, *see* Particle in a box
Bulk modulus, *see* Elastic moduli

Carrier concentrations, extrinsic semiconductor, 197, 214, 215, 216, 217, 219, 221, 222, 223, 224, 225, 226, 227, 228, 229, 230, 231, 232, 234, 235, 241, 246, 248
  intrinsic semiconductor, 197, 204, 205, 206, 207, 208, 209, 210, 211, 212, 213, 214, 215, 216, 221, 234, 235, 241, 242
Carrier mobility, *see* Mobility
Cavity radiation, *see* Black body
Certainty, in information theory, 256, 257
Change of thermodynamic state, *see* Thermodynamic change of state

269

Characteristic temperature, *see* Temperature
Chemical equilibrium, *see* Equilibrium potential, 68
Classical approximation, *see* Maxwell–Boltzmann tail
Collision cross-section, 153, 154, 157
Collisional relaxation time, 144, 145, 146, 148, 149, 165, 234, 235, 236, 239
Collisions, electron–electron, 146
   electron–impurity atom, 157, 158, 168, 169, 170, 183, 185
   electron-ionized impurity atom in a semiconductor, 237, 238, 243, 248
   electron–phonon, 133, 149, 150, 152, 156, 157, 158, 164, 166, 167, 169, 172, 175, 181, 183, 184, 185, 188, 237, 238
   hole–phonon, *see* Electron–phonon
   phonon–impurity atom, 172, 237, 238, 248
   phonon–phonon, 130, 172, 174, 175, 176, 178
Communication channel, 255, 256, 257
   entropy, 263
Compensation, *see* Extrinsic semiconductor
Concentration of particles, *see* Carrier concentration, Valence electron gas
Conduction band, *see* Band theory
   electrons, *see* Valence electron gas
Conservation laws, 262
Constant pressure heat capacity, 104, 105
Constant volume heat capacity, monatomic Maxwell–Boltzmann gas, 70, 100
   Debye theory, 121, 125, 126, 127, 128, 130, 132, 167, 176, 177
   Einstein theory, 113
   Maxwell–Boltzmann plasma, 77
   relationship with thermodynamic energy, 4, 70, 77, 121
   total, of a solid, 105, 131
   valence electron gas in a metal, 85, 86, 100, 101, 102, 103, 105, 130, 131, 165
Contact potential in semiconductor, 253
Continuous distribution of translational energy levels, *see* Particle in a box

Conversion factor, *see* Energy conversion factor
Critical point, 63
Current density, in a metal, 139, 140, 141, 142, 144, 145
   in a semiconductor, 233, 251
Crystal, *see* Solid
   momentum, *see* Momentum

Debye theory, continuous distribution of frequencies, 119, 120, 123, 126, 128, 130
   cut-off frequency, 118, 119, 123, 125, 126, 128, 150, 153, 160
   distribution of frequencies, 115, 120, 130
   function, 120, 121, 122
   heat capacity, *see* Constant volume heat capacity
   of a solid, *see* Lattice
   standing thermal waves, 115, 171, 172
   T^3 low temperature law, 105, 113, 126, 128, 167, 178
   temperature, *see* Temperature
Debye–Hückel screening, 191, 238
Degeneracy criterion, *see* Particle in a box
   translational level, *see* Particle in a box
Degenerate electron gas, *see* Valence electron gas
Degree of ionization, in a semiconductor, *see* Extrinsic semiconductor
   in a Maxwell–Boltzmann plasma, *see* Ionization equilibrium
Density of states function, 23, 60, 83, 93, 198, 200, 203
Diameter, collision, *see* Collision cross-section
Diatomic molecule, rotation of, 71
   vibration of, 71, 106
Diffusion, carriers in a semiconductor, 251
   velocity in a semiconductor, 252
Diffusion coefficients, and Fick's law, 137, 139, 141, 251
   table of, 253
Disorder, *see* Entropy
Distribution functions, *see* Bose–Einstein, Fermi–Dirac, Maxwell–Boltzmann distribution functions
   continuous, *see* Particle in a box
   hole, *see* Hole distribution function

Index

Donor atom, *see* Extrinsic semiconductor
Drift velocity in an electric field, 142, 144, 145, 151, 233, 234, 236, 237, 252
Dulong–Petit law, 105, 125
Dynamical model of a lattice, *see* Lattice

Effective mass, 133, 140, 144, 175, 207, 208, 209, 210, 211, 213, 234
Einstein function, 121, 122
  heat capacity theory, *see* Constant volume heat capacity
  theory of a solid, *see* Lattice
Elastic moduli, 114, 126, 130, 174
  thermal waves, *see* Debye theory
Electrical conductivity, definition and Ohm's law, 85, 139, 141, 145, 146, 233
  of an extrinsic semiconductor, 194, 234, 243, 244, 245, 246
  of an intrinsic semiconductor, 194, 234, 241, 242, 243
  of a metal, 89, 114, 141, 142, 144, 148, 152, 153, 154, 155, 158, 160, 180, 181, 183, 184, 185, 186, 187, 194, 234, 239, 240
  phonon component, in a metal, 158, 159, 183, 184, 185, 186
  pressure variation of, in a metal, 160, 161
  residual, in a metal, 157, 158, 159, 183, 185, 186, 194, 235, 238, 239
  table of values, in a metal, 159
  in a semiconductor, 245
  temperature variation of, in an extrinsic semiconductor, 243, 244, 245
  in an intrinsic semiconductor, 241, 242, 243
  in a metal, 141, 155, 156, 157, 159, 160, 183
  variation with doping, in an extrinsic semiconductor, 245, 246, 247
Electrical resistivity, *see* Electrical conductivity
Electromagnetic radiation, *see* Black body
Electron charge, numerical value of, 264
  collisions, *see* Collisions
  concentration, *see* Carrier concentration, Valence electron gas
  conduction, *see* Valence electron gas
  energy states of an atom, 72, 73, 74, 78, 80, 196
  gas, *see* Valence electron gas
  mass, numerical value of, 264
  mean free path, *see* Mean free path
  mobility, *see* Mobility
  partition function, *see* Partition function
  spin quantum states, 75, 81, 85, 87, 91, 98, 198, 207, 218
  volt, *see* Energy conversion factor
Energy, average thermal, 70, 92, 97, 101, 140, 147, 148, 150, 162, 165, 235 236
  band, *see* Band theory
  conservation of, *see* Conservation laws
  conversion factor, joules to electron volts, 264
  Fermi, *see* Fermi energy
  gap, *see* Band theory
  in transit, *see* Thermodynamic interaction
  level, *see* Particle in a box
  quantization, *see* Particle in a box
  rotational, *see* Diatomic molecule
  thermodynamic, *see* Thermodynamic energy
  vibrational, *see* Diatomic molecule
  zero point, *see* Zero point energy
Entropy, increase of, 261
  non-conservation of, 262
  partition function and, *see* Partition function
  randomness and, 261, 262
  Second Law and, *see* Second Law
  statistical definition of, 54, 65, 259, 260, 262, 263
  Third Law and, 54, 260
Equilibrium, condition for thermodynamic, 7, 8, 64, 68
  chemical, 8, 65, 68, 78
  distribution function, *see* Distribution function
  ionization, *see* Extrinsic semiconductor, Ionization equilibrium
  mechanical, 7, 65, 68, 107, 137
  thermal, 7, 64, 68, 107, 128, 130, 137, 140, 176, 253
  thermodynamic, 7, 43, 51, 53, 64, 65, 66, 68, 137, 138, 155, 253, 260
Exclusion principle, *see* Pauli exclusion principle
Extrinsic carrier concentration, *see* Carrier concentration

Extrinsic semiconductor, carrier concentration in, *see* Carrier concentration
  charge neutrality in, 217, 223, 225, 246
  compensation in, 222, 223, 246
  definition of, 193, 194, 195
  Fermi level in, *see* Fermi level
  impurity atom levels in, 195, 196
  intrinsic temperature range, 229, 230, 244, 245
  ionization of impurity atoms, 195, 213, 215, 217, 218, 219, 220, 221, 222, 225, 226, 228, 229, 230, 231, 232, 239, 243, 245, 246
  ionization energy of impurity atoms, 195 196, 197, 198, 209, 220, 241
  Law of Mass Action, 214, 215, 216, 223, 224, 225, 246
  level of doping in, 194, 195, 196, 221, 223, 225, 228, 229, 231, 232, 238, 239, 243, 246
  saturation region, 219, 228, 229, 230, 231, 232, 243, 244, 245

Fermi energy, in extrinsic semiconductor, 215, 216, 218, 219, 220, 222, 223, 224, 225, 226, 228, 229, 253
  in intrinsic semiconductor, 202, 203, 206, 207, 208, 209, 211, 222, 223, 224, 253
  in a metal, 83, 86, 87, 88, 89, 90, 91, 92, 93, 94, 95, 96, 97, 98, 99, 101, 102, 132, 140, 143, 146, 147, 148, 150, 151, 155, 165, 167, 196, 209, 210, 235, 237, 260
  relation with valence electron concentration in a metal, 91, 96, 147, 209
  rule for finding, 86, 87, 88, 89, 101, 210, 211, 222
  temperature variation of in a semiconductor, 166, 167, 168, 204, 211, 222, 223, 224, 227
Fermi temperature, *see* Temperature
Fermions, 33, 39, 40, 50, 81, 101, 199
Fermi–Dirac distribution function, 49, 55, 58, 86, 87, 89, 90, 92, 93, 96, 98, 199, 201, 202, 203, 204, 210, 211, 212, 218, 226, 227
  partition function, *see* Partition function
  statistics, 23, 33, 38, 41, 42, 44, 46, 47, 48, 50, 85, 86, 90, 100, 140, 149, 199, 201, 202, 203
  valence electron gas, *see* Valence electron gas
Fick's law of diffusion, *see* Diffusion coefficient
Films, thin, *see* Thin films
First Law of Thermodynamics, 5, 51, 64, 249
  differential form of, 6, 66
  integrated form of, 6, 249
  thermodynamic energy and, 5
Flux of electric charge, *see* Current density
  of thermal energy, *see* Heat transfer
Forbidden band, *see* Band theory
Fourier's law of heat conduction, *see* Thermal conductivity
Free electron gas, *see* Valence electron gas

Gas constant, universal, *see* Universal gas constant
Gibbs' function, 78
Gibbs–Dalton law, 55
Gradient of intensive property, *see* Transport processes
Ground state, *see* Energy states of an atom
Group velocity, phonon, 117
  photon, 116

Harmonic oscillator, amplitude of, 109, 110, 112, 152, 153
  average kinetic energy of, 111
  average potential energy of, 111
  average thermal energy of, 152, 153
  definition of, 107
  kinetic energy of, 109, 152
  potential energy of, 109, 152, 173
  table of functions of, 122
  total energy of, 110, 112, 152
Heat capacity, *see* Constant volume heat capacity
Heat transfer, as an interaction, 2, 4, 5, 6, 7, 8, 65, 66, 100, 101, 104, 137, 249
  irreversible, 9, 11
  microscopic aspect of, 51, 106, 162
  reversible, 9, 11, 53, 54, 261
  sign convention, 5
Heisenberg uncertainty principle, 116

Index

Hole distribution function in semiconductor, 200
Holes, *see* Extrinsic, Intrinsic semiconductor
Hooke's law for a spring, 108, 114, 172, 173

Ideal gas, *see* Thermally perfect gas
Identification of α and β, *see* Undetermined multipliers
Imperfections in a lattice, *see* Lattice
Impurity atoms, *see* Collisions, Electrical conductivity, Extrinsic and Intrinsic semiconductor, Thermal conductivity
Impurity coefficient of a lattice, 169, 170, 186
Independent particles, *see* Thermally perfect gas
Indistinguishable particles, 14, 15, 28, 33, 35, 39
Information, in a message, 255, 256, 257, 258
 per information symbol, 257, 258, 259
Information receiver, 255
 source, 255, 256
 symbols, 255, 256
Insulator, 86, 100, 101, 103, 113, 132, 133, 180, 182, 190, 191, 193, 194, 195, 196, 235, 247
Interaction energy, *see* Thermally perfect gas
Interactions, *see* Thermodynamic interactions
Intrinsic carrier concentration, *see* Carrier concentration
 range of extrinsic semiconductor, *see* Extrinsic semiconductor
Intrinsic semiconductor, carrier concentration in, *see* Carrier concentration
 charge neutrality, 197, 208, 209, 210, 215
 defined, 193, 194
 Fermi level in, *see* Fermi level
 Law of Mass Action, 208, 209, 215
 residual impurities in, 238, 239
Ion, 73, 74, 75, 76
Ionization energy, of atom in Maxwell–Boltzmann plasma, 76, 78, 79
 of impurities in semiconductor, *see* Extrinsic semiconductor

Ionization equilibrium of Maxwell–Boltzmann plasma (Saha equation), 76, 77, 78, 209, 220
Irreversible process, *see* Thermodynamic process
Isolated thermodynamic system, *see* Thermodynamic system

Joule, conversion factor for, *see* Energy conversion factor
Joule heat, 143, 151, 232, 243, 249
Junction, p–n, *see* p–n junction

Kelvin temperature, *see* Temperature
Kilogram mole, *see* Mole
Kinetic theory, 61, 138, 143, 154, 161, 162, 163

Lagrange's method of undetermined multipliers, 45, 47
Lattice, dynamical model of, 80, 102, 106, 113, 114, 124, 125, 126, 127, 128, 129, 130, 151, 152, 170, 173, 174, 193
 Debye model of, 106, 113, 114, 115, 116, 117, 119, 120, 121, 123, 125, 129, 154, 160, 170, 171, 172, 173, 174, 175
 Einstein model of, 106, 113, 126, 154, 173, 174, 175
 heat capacity, *see* Constant volume heat capacity
 imperfections, dynamic, 142, 143, 146, 148, 149
 static, 142, 143, 146, 148, 149
 nuclear structure of, 105, 106, 113, 124, 126, 127, 128, 129, 152, 173
 thermal conductivity, *see* Thermal conductivity
Law of Mass Action, *see* Extrinsic, Intrinsic semiconductor
Localized phonon, 116, 117
Lorenz number, in a metal, 185, 186, 187
 in a semiconductor, 247
 table of values for metals, 187

Macroscopic description of gas, *see* Thermally perfect gas
Macrostate, 15, 18, 21, 24, 25, 27, 29, 30, 32, 33, 36, 37, 38, 39, 40, 42, 43, 44, 48, 50, 117, 118, 119, 259, 260, 261
 equations of constraint of, 16, 28, 29, 33, 36, 40, 45, 46, 47, 56, 137

most probable, 15, 42, 43, 44, 45, 48, 49, 50, 51, 53, 55, 137, 261
occupation index, 48, 49, 50, 87
occupation numbers, 45, 48, 49, 94, 95
probability of, 30, 31, 43
Mass, effective, *see* Effective mass
electron, *see* Electron mass
Mass Action, law of, *see* Extrinsic, Intrinsic semiconductor
Mass transfer, 2, 5, 65, 116
Material oscillator, *see* Lattice
Matthiesen's rule, 158, 169, 172
Maxwellian distribution of speeds and energies, 23, 61, 176
Maxwell–Boltzmann, distribution function, 49, 55, 59, 61, 90, 201, 202, 203, 204, 205, 207, 212, 235
monatomic gas, 69, 92, 97, 101
partition function, *see* Partition function
plasma, 64, 69, 74, 75, 90
statistics, 33, 41, 42, 46, 47, 48, 50, 57, 58, 61, 63, 69, 74, 85, 90, 100, 140, 201, 202, 214, 237
'tail' as common limiting statistical form, 41, 42, 49, 57, 58, 62, 90, 140, 201, 202, 203, 205, 207, 212, 214, 235, 247
Maximum term method, 42
Mean free path, carriers in a semiconductor, 236, 241
conduction electrons in a metal, 143, 148, 149, 151, 154, 155, 156, 157, 158, 164, 165, 166, 167, 168, 169, 183, 186, 187, 234, 235, 236, 237
phonons in a solid, 164, 176, 178, 179, 181, 183
Mean square speed, *see* Speed
Mechanical equilibrium, *see* Equilibrium
Message, in information theory, 255, 256
Metals, *see* Constant volume heat capacity, Electrical conductivity, Thermal conductivity Valence electron gas
Method of undetermined multipliers, *see* Lagrange's method
Microscopic description, of Thermally perfect gas, *see* Thermally perfect gas
of heat transfer, *see* Heat transfer
of work transfer, *see* Work transfer
Microstate, 27, 29, 30, 32, 33, 36, 37, 38, 39, 40, 42, 43, 44, 45, 48, 50, 259, 260, 261, 263
probability of, 30, 31, 34, 35, 36, 37, 40, 43, 262, 263
Mobility, of carriers in an extrinsic semiconductor, 239, 240, 241, 246
of carriers in an intrinsic semiconductor, 238, 239, 240, 241, 242, 243
of electrons in a metal, 239, 240
of electrons in a semiconductor, 233, 234, 236, 237, 238, 240, 241
table of values of, 240
temperature variation, for carriers in a semiconductor, 236, 237, 238, 239, 242, 243
Molar values for a solid, 71, 125, 266, 267
Mole, definition of, 105, 265
Molecular weight, *see* Atomic weight
Molecule, *see* Diatomic molecule
Momentum, crystal, 175
phonon, 116, 117, 175, 187, 188
photon, 116
valence electron, 150, 187, 188
Monatomic gas, *see* Maxwell–Boltzmann
Most probable macrostate, *see* Macrostate

n–p product, *see* Law of Mass Action
n-processes (normal processes) 75
Noble metals, 80, 92, 154, 193, 195
Non-degenerate electron gas, *see* Valence electron gas
Non-equilibrium particle distribution function, *see* Boltzmann transport equation
Non-equilibrium thermodynamics (thermodynamics of the steady state), 137, 138
Normal modes of oscillation, *see* Debye theory
Number density, *see* Concentration of particles
Number, of atoms in a solid, 115, 116, 117, 118, 119, 120, 121, 123, 125, 126, 127, 131, 133, 134, 154, 156, 157, 160, 163, 166, 177, 178, 179, 195, 213
calculation of, 265, 266, 267

Index

of impurity atoms in a metal, 157, 168, 169
of particles and partition function, *see* Partition function

Occupation index, *see* Macrostate
Occupation number, *see* Macrostate
Ohm's law, *see* Electrical conductivity
Open thermodynamic system, *see* Thermodynamic system
Oscillator, anharmonic, *see* Anharmonic oscillator
harmonic, *see* Harmonic oscillator

p–n junction, 64, 137, 232, 233, 249, 253
Fermi level alignment at, 253
Particle in a box, continuous distribution of energy levels, 23, 24, 31, 60, 71, 72, 92, 93, 98, 119, 198, 199, 210
degeneracy criterion, 62, 63, 90, 214
degeneracy or statistical weight of energy levels, 19, 20, 21, 23, 25, 27, 28, 30, 34, 38, 39, 83, 260, 261
difference between energy state and energy level, 20, 24, 25, 26, 27, 32, 34
distribution function for continuous energy levels, *see* Bose–Einstein, Fermi–Dirac, Maxwell–Boltzmann distribution function
potential barrier for, 17, 84, 85, 192
potential energy of, 14, 15, 16, 52, 53, 83, 84, 85, 91, 190, 191, 192
quantal boundary conditions for, *see* Boundary conditions
quantization of translational energy, 15, 17, 24
standing waves of, 82, 83, 115
structure of, 15, 16, 24, 52, 69, 71, 75, 100, 259
translational or kinetic energy of, 15, 17, 18, 19, 20, 21, 22, 24, 27, 38, 39, 69, 71, 81, 84, 85, 101, 193, 235, 259, 260, 261
translational energy ('eigen') states, 16, 17, 18, 19, 20, 21, 22, 23, 25, 26, 27, 28, 29, 30, 31, 33, 34, 43, 51, 52, 53, 74, 81, 82, 83, 84, 88, 90, 91, 93, 101, 115, 143, 150, 151, 165, 191, 193, 259, 260

translational quantum numbers of, *see* Quantum numbers
translational wave functions ('eigenfunctions') of, 17, 18, 21, 38, 82, 115
Partition function, 55, 56, 57, 63, 69, 104, 105
electronic, 73, 74, 75, 76, 78, 206
entropy and, 56, 59, 70
generalized Bose–Einstein and Fermi–Dirac, 56, 58, 59, 62, 63, 98, 105, 205, 206, 261
Maxwell–Boltzmann, 58, 59, 60, 61, 69, 70, 72, 73, 75, 105, 205, 206
number of particles and, 56, 59, 206
phonon gas, 119, 120, 121
pressure and, 56, 57, 59, 70, 121
thermodynamic energy and, 56, 59, 120
translational, 61, 69, 206
valence electron gas, 98
Pauli exclusion principle, 24, 31, 33, 38, 40, 50, 81, 85, 87, 91, 92, 93, 94, 101, 118, 150, 191, 196, 198, 199, 218, 260
Perfect gas, *see* Thermally perfect gas
Periodic, boundary conditions, *see* Boundary conditions
Periodic potential, *see* Band theory
Phonons, boundary scattering of, *see* Boundary scattering
collisions between, *see* Collisions
defined, 115, 116
distribution function, *see* Bose–Einstein distribution function
mean free path, *see* Mean free path
momentum, *see* Momentum
number in a solid, *see* Number of phonons
production of, in a heated solid, 117, 118, 133, 134, 135, 177, 178
quantum states, 118, 119
Phonon gas, *see* Lattice
Photon, defined, 112, 113
Physical constants, numerical values of, 264
Planck's constant, numerical value of, 264
Plasma, *see* Maxwell–Boltzmann plasma
Potential barrier, *see* Particle in a box
Potential energy, of electron in a metal, *see* Valence electron gas
of electron in a semiconductor, *see* Band theory

Pressure, microscopic description of, 52, 54
  of phonon gas in a solid, 117, 121, 123
  of valence electron gas in a metal, 97, 98, 99
  partition function and, *see* Partition function
Probability, of a macrostate, *see* Macrostate
  of a microstate, *see* Microstate
  of a symbol in information theory, 257, 258, 259
  of the outcome of an event, 256, 257, 262
Process, thermodynamic, *see* Thermodynamic process
Production of entropy, *see* Entropy
Process, n-type and u-type, *see* n-type, u-type processes

Quantum numbers, electronic, 72
  spin, 81, 82, 85, 88, 98
  translational, 17, 19, 20, 22, 23, 24, 72
Quantum number space, 21, 91
Quantum oscillator, *see* Harmonic oscillator
Quantum restriction, *see* Pauli exclusion principle
Quasifree electron, *see* Band theory
Quasistatic process, *see* Thermodynamic process

Radiation, black body, *see* Black body
  oscillator, *see* Black body
Randomness, *see* Entropy
Relaxation time, *see* Collisional relaxation time
Residual resistivity, *see* Electrical conductivity
Resistance, temperature coefficient of, 156, 159, 242, 243
Resistivity, electrical, *see* Electrical conductivity
  thermal, *see* Thermal conductivity
Rest mass, 55, 106, 116, 117, 118, 174
Reversible process, *see* Thermodynamic process
Root mean square speed, *see* Speed
Rotational energy, *see* Diatomic molecule

Sackur–Tetrode equation, 70

Saha equation, *see* Ionization equilibrium
Scattering processes, *see* Collisions
Schrödinger wave equation, 13, 16, 17, 27, 191
Screened coulomb potential, *see* Debye–Hückel screening
Second Law of Thermodynamics, 5, 10, 51, 64, 67, 261
  entropy and, 11, 12, 54, 261
Shannon's formula, 259
Single particle translational energy states, *see* Valence electron gas
Solid, different types of crystal binding, 80
Solid, valence electron translational energy states of, *see* Valence electron gas
Sommerfeld model, *see* Valence electron gas
Sound speed, *see* Speed
Specific heat, *see* Constant volume heat capacity
Speed, Fermi, 147, 148, 149, 151, 154, 155, 157, 165, 166, 169, 234, 235, 236, 240
  mean square thermal, 140, 236
  of light, 116
  numerical value of, 264
  of sound (phonon speed), 116, 123, 126, 130, 174, 176, 177, 178, 179
  root mean square thermal, 140, 148, 151, 164, 176, 236, 240
Spin, electron, *see* Electron
  quantum number, *see* Quantum number
  quantum states, *see* Electron
Standing wave, *see* Debye theory, Particle in a box
State, thermodynamic, *see* Thermodynamic state
Statistical weight or degeneracy, *see* Particle in a box
Stirling's theorem, 44, 45, 258
Strain gauge, 161
Symmetrical approximation in Fermi–Dirac statistics, 89, 210, 211

Temperature, absolute or Kelvin, 4
  Debye, 123, 124, 125, 126, 128, 129, 130, 131, 141, 150, 153, 154, 155, 156, 157, 160, 161, 164, 166, 167, 168, 169, 170, 177, 178, 179, 183, 184, 185, 186, 187, 194, 237, 238, 239, 243, 248

Index

Fermi, 99, 100
 table of melting, 212
 transition, for extrinsic semiconductor 230, 231, 232, 245
Thermal conductance, 250
Thermal conductivity, definition and Fourier's law, 137, 139, 163, 171, 249
 kinetic theory relationship with heat capacity, 161, 163
 lattice, boundary scattering phonon component, 179
 lattice, impurity atom component, 180, 181, 182
  phonon component, 177, 178, 179, 180, 181
  total, 169, 180, 247, 248
 of a semiconductor, 247, 248, 249 250, 251
 of valence electron gas, boundary scattering electron component, 167, 183
  impurity atom component, 168, 169, 181, 183, 186
  phonon component, 89, 164, 165, 166, 167, 168, 169, 170, 181, 183, 184, 186
  total, 169, 170, 184, 186
 tables of values, metal, 170
  non-metal, 182
  semiconductor, 248
 temperature variation of lattice component, 177, 178, 179, 180, 182, 248
 total, of a solid, 180, 181, 194
Thermal excitation, 102, 193, 194, 195, 197, 199, 212, 214, 215, 227, 229
 equation of state, see Thermally perfect gas
 expansion, 130, 160, 173
 radiation, see Black body
 resistance, 250
 resistivity, see Thermal conductivity
Thermally perfect gas, description of, 13, 14, 15, 50, 55, 64
 equation of state, definition of, 10, 54, 55
  Bose–Einstein phonon gas in a solid, 55, 123, 125, 127
  Maxwell–Boltzmann monatomic gas, 54, 70
  Maxwell–Boltzmann plasma, 75, 76, 77
 Valence electron gas in a metal, 54
 independent particles of, 15, 24, 33, 50, 85, 116, 259
 interaction energy between particles of, 13, 24, 33, 50, 74, 75, 85, 116, 259
 macroscopic description of, 14, 16, 50, 51
 microscopic description of, 15, 16, 50 51, 85
Thermistor, 242
Thermodynamic change of state (or process), defined, 2, 3
 irreversible, 9, 10, 12
 reversible or quasistatic, 9, 10, 12, 51, 137
 disequilibrium, 8, 9, 137, 138, 155
 equilibrium, see Equilibrium
 interaction, see Heat, Mass, Work transfer
 probability, see Macrostate, Microstate
 properties, defined, 3
  examples of, 4, 7, 78, 99, 104, 105, 262
 state, defined, 3, 14
 surroundings, defined, 2
  example of, 4
Thermodynamic system, closed, 2, 3, 7, 14, 48, 51, 53, 64, 66, 137, 249, 259
 defined, 1
 example of, 1, 4, 104, 249
 isolated, 2, 66, 67, 68
 open, 8, 65, 66
Thermodynamic system boundary, defined, 1
 example of, 1, 2, 4
Thermodynamics, defined, 1, 2
Thermoelectricity, 64, 195
Thin films, electrical conductivity of, 158, 167
 thermal conductivity of, 167
Third Law of Thermodynamics, see Entropy
Transistor, 64, 214, 232, 249, 250, 251
Translational energy, see Particle in a box
 partition function, see Partition function
 quantum number, see Quantum numbers
 wave function, see Particle in a box

Transport coefficient, *see* Transport processes
Transport processes, 84, 117, 137, 138, 139, 147, 155, 171, 249, 251, 252
and equilibrium approximation theory of, 138, 143, 249
Tunnel diode, 195

u-processes (umklapp processes), 175, 176, 178
Uncertainty, in information theory, 256, 257, 262
Undetermined multiplier α, 47, 86, 118, 119, 206, 253
β, 47, 49, 53, 59, 61
Universal gas constant, in solids, 71, 105
numerical value of, 264

Valence band, in a metal, 81, 82, 83, 84, 85, 88, 92, 93, 95, 191, 195, 196
in a semiconductor, *see* Band theory
Valence electron gas, Bloch potential for, *see* Band theory
free electrons of, 63, 80, 81, 82, 83, 84, 85, 86, 87, 88, 89, 90, 92, 96, 97, 99, 100, 101, 102, 103, 104, 105, 130, 132, 139, 140, 143, 145, 148, 150, 164, 165, 190, 191, 201, 259
non-degenerate conduction electrons of, 69, 80, 89, 90, 96, 99, 101, 102, 131, 140, 141, 142, 143, 146, 148, 149, 150, 151, 153, 164, 165, 181, 182, 183, 185, 188, 201, 237
number of electrons in a metal, 91, 92, 93, 94, 96, 98, 102, 146, 154, 169, 208, 209, 213, 234, 240

partition function of, *see* Partition function
pressure of, in a metal, 97, 98, 99
single particle translational energy states of, in a metal, 38, 39, 40, 81, 82, 86, 87, 88, 89, 92, 93, 94, 95, 103
in a semiconductor, 198, 199, 200, 201, 203, 204, 207, 218, 260
Valency, defined, 266
of various metals, 92, 100, 132, 133, 154, 267
Varistor, 243
Velocity, drift, *see* Drift velocity
Vibrational energy, of a lattice, *see* Lattice
of a molecule, *see* Diatomic molecule

Wave equation, *see* Schrödinger wave equation
function, *see* Particle in a box
packet, *see* Localized phonon
standing, *see* Standing wave
Wiedemann–Franz–Lorenz law, for carrier in a semiconductor, 246
for electrons in a metal, 85, 141, 150, 182, 185, 186, 187, 188
Work transfer, as interaction, 2, 4, 5, 6, 7, 11, 65, 66, 137
displacement, 8, 10, 53, 65, 67
irreversible, 9, 10
microscopic aspect of, 51, 52
reversible, 9, 10, 51, 53
sign convention for, 5

Zero point energy of vibration, 106
Zero'th Law of Thermodynamics, 4

TJ
265
M23

MAR 8 1973